Petri Nets

Guanjun Liu

Petri Nets

Theoretical Models and Analysis Methods for
Concurrent Systems

 Springer

Guanjun Liu (iD)
Department of Computer Science
Tongji University
Shanghai, China

ISBN 978-981-19-6311-7   ISBN 978-981-19-6309-4   (eBook)
https://doi.org/10.1007/978-981-19-6309-4

This Springer imprint is published by the registered company Springer Nature Singapore Pte Ltd.
The registered company address is: 152 Beach Road, #21-01/04 Gateway East, Singapore 189721,
Singapore

Preface

Petri nets, as the first true concurrent model, have been of good vitality since Dr. Carl Adam Petri pioneered them sixty years ago. Such a good vitality benefits from the outstanding work of Dr. Petri and other scientists and technicians in this field, indicating that Petri nets (including high-level extensions) have many advantages in modelling concurrent systems as well as in helping us understand, analyse, and improve these concurrent systems.

Concurrency is a universal phenomenon in our physical world. Even when an organisation or a team performs a task, concurrent operations are often taken in order to increase efficiency. In the current scientific times, lots of concurrent systems serve our work and life, but they must provide a correct and reliable service, or our safety and even lives are threatened. However, designing such a system is not an easy thing, because a number of concurrent, interactive, and even real-time actions are involved in its running so that its logical structures and dynamic behaviours are very complex. In consequence, we need one or multiple suitable models to help us understand and analyse the logical structures and dynamic behaviours of such a complex system, while Petri nets have been well used just due to their ability on simulating concurrence, interaction, real time, etc.

Purpose

In addition to a part of my research results, what this book focuses on is the basic theory of Petri nets as well as the Petri-nets-based model checking methods, so that it is suitable not only for senior researchers, but also for beginners aspiring to this field. The biggest feature of this book, I think, is a relatively systematic combination of the Petri net theory and the Petri-nets-based model checking methods. The Petri net theory can help readers systematically understand the nature of concurrence, interaction, and real time. The model checking methods are important practical techniques employed for checking errors in concurrent systems. So far, only few books combine them together.

Prerequisites

The Petri net theory and model checking methods have their roots in mathematical foundations such as discrete mathematics, propositional logic, automata theory, data structures, graph theory, and algorithms. It is expected that readers are familiar with the basic knowledge of these aspects when starting with this book.

Content

As for the Petri net theory, this book involves the basic knowledge of elementary net systems, including their interleaving semantics and concurrency semantics (Chap. 1), some structure concepts (Chap. 2), some subclasses with special structures (Chaps. 3 and 4), and some basic properties such as reachability, liveness, and deadlock (Chaps. 1–4). It also involves four high-level Petri nets: Knowledge-oriented Petri nets (Chap. 6), Petri nets with insecure places (Chap. 7), time Petri nets (Chap. 8), and plain time Petri nets with priorities (Chap. 10), focusing on different application fields. As for the model checking methods, this book involves computation tree logic (Chap. 5), computation tree logic of knowledge (Chap. 6), timed computation tree logic (Chaps. 9 and 10), as well as the Petri-nets-based checking methods for them. The basic principle of the reduced ordered binary decision diagram is also introduced in order to compress the state space used in these model checking procedures (Chaps. 5 and 6). Besides the above contents, this book also presents time-soundness for time Petri nets (Chap. 8) and secure bisimulation for Petri nets with insecure places (Chap. 7). They are both based on the bisimulation theory pioneered by Dr. Robin Milner, and thus this theory is also introduced briefly in this book.

Acknowledgements

I would like to thank all my colleagues for their hard work in fighting against COVID-19. COVID-19 forced me to stay at home; but on the other hand, I had enough time to write this book. I thank the National Nature Science Foundation of China (Nos. 62172299 and 62032019) and the Alexander von Humboldt Foundation of Germany. They supported my research so that a part of my research results can be written in this book. At last, I thank my wife, my daughters, and my mother for their selfless support.

Shanghai, China   Guanjun Liu
July 2022

Contents

Chapter 1
Elementary Net Systems

1.1 Net Diagram and Semantics

This section first introduces some basic notions and notations of elementary net systems, and then describes their interleaving semantics and concurrent semantics, respectively. More details can be seen in [1–3].

The following notations are used in the whole book. $\mathbb{N} \triangleq \{0, 1, \ldots\}$ represents the set of all nonnegative integers and $\mathbb{N}^+ \triangleq \{1, 2, \ldots\}$ stands for the set of all positive integers. Given $m \in \mathbb{N}$, $\mathbb{N}_m \triangleq \{0, \ldots, m\}$ represents the set of integers from 0 to m and $\mathbb{N}_m^+ \triangleq \{1, \ldots, m\}$ denotes the set of integers from 1 to m. Obviously, $\mathbb{N}_m^+ = \emptyset$ when $m = 0$. Sometimes, we use $\mathbb{N}_1 = \{0, 1\}$ to represent the set $\{\textbf{false}, \textbf{true}\}$ of truth values in the propositional logic.

1.1.1 Net and Net System

1.1.1.1 Net Diagram

Definition 1.1 (*Net*) A *net* is a 3-tuple $N \triangleq (P, T, F)$ where P is a finite set of *places*, T is a finite set of *transitions*, $F \subseteq (P \times T) \cup (T \times P)$ is a *flow relation*, $P \cup T \neq \emptyset$, and $P \cap T = \emptyset$.

Example 1.1 Figure 1.1 illustrates a net that is represented by a directed bipartite graph where the dots in cycles need to be neglected temporally. Its place set, transition set and flow relation are formally defined as follows:

$$- \ P = \bigcup_{j=1}^{5} \{p_{j,1}, \ p_{j,2}, \ p_{j,3}, \ r_j\},$$

© The Author(s), under exclusive license to Springer Nature Singapore Pte Ltd. 2022
G. Liu, *Petri Nets*,
https://doi.org/10.1007/978-981-19-6309-4_1

Fig. 1.1 A Petri net modelling the dining philosophers problem that was first proposed by Dr. Edsger Wybe Dijkstra [4] and later formally described by Dr. Tony Hoare [5]. It reveals the phenomenon that multiple processes share finitely many resources and thus deadlock or starvation can occur if these resources are allocated improperly. Here is a simplified version of 5 philosophers chosen from [6]

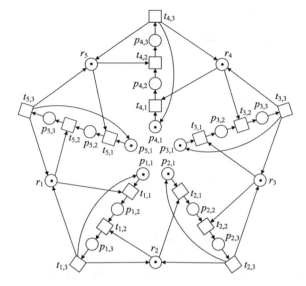

$$- \; T \; = \; \bigcup_{j=1}^{5} \{t_{j,1}, \; t_{j,2}, \; t_{j,3}\},$$

$$- \; F \; = \; \bigcup_{j=1}^{5} \{(p_{j,1}, \; t_{j,1}), \; (t_{j,1}, \; p_{j,2}), \; (p_{j,2}, \; t_{j,2}), \; (t_{j,2}, \; p_{j,3}), \; (p_{j,3}, \; t_{j,1}),$$

$$(r_j, \; t_{j,1}), \; (t_{j,3}, \; r_j), \; (r_{(j \bmod 5)+1}, \; t_{j,2}), \; (t_{j,3}, \; r_{(j \bmod 5)+1})\} \, .$$

From this example it can be seen that a net can be viewed as a directed bipartite graph in which cycles represent places, boxes represent transitions, and arcs represent the flow relation.

The net in Fig. 1.1 models the dining philosophers problem. Five philosophers sit around a circular table and there is a chopstick between any two adjacent philosophers. A philosopher first holds his left chopstick and then his right one. After holding the two chopsticks, the philosopher can eat. After finishing eating, the philosopher returns the thinking state again and releases the two chopsticks. The places r_j, $j \in \mathbb{N}_5^+$, stand for five chopsticks. The place $p_{j,1}$ stands for the thinking state of the j-th philosopher, the place $p_{j,2}$ stands for the state where the j-th philosopher has holden the left chopstick, and the place $p_{j,3}$ stands for the state where the j-th philosopher has also holden the right chopstick and is eating. The flow relation of the net represents the logical (temporal) relation of states of each philosopher:

thinking → holding left chopstick → holding two chopsticks and eating → thinking

and the logical relation of states of each chopstick:

idle → working → idle.

A net generally represents the static logical structure of a concurrent system but cannot yet simulate its dynamic behaviours (runs).

Just according to the net in Fig. 1.1, we cannot know which state every philosopher is at or who is holding a chopstick. In other words, such a net cannot yet reflect the dynamic behaviours of a system since different *places* just stand for different *conceptive local states* of the system, but so far a *concrete global state* of the system has not been represented. Therefore, concrete global states should be defined on the basis of these conceptive local states, representing which conceptive local states (places) are active at a given concrete global state.

The changes of concrete global states, representing the dynamic behaviours of a system, can be caused by those *transitions* that stand for different actions or events.

Next, concrete global states are defined on the basis of conceptive local states. A concrete global state is generally called a *marking* in the Petri net theory.

1.1.1.2 Marking and Net System

Definition 1.2 (*Marking*) A *marking* of a net $N = (P, T, F)$ is a mapping M: $P \to \mathbb{N}$.

Given a marking M and a place p, the value of $M(p)$ represents the situation of the place p at the current concrete global state M. For instance, if a place p stands for the eating state of some philosopher, then $M(p) = 1$ means that the philosopher is eating at the current global state, while $M(p) = 0$ means the philosopher is not eating at this global state. For another instance, if a place p stands for a class of machines, then the value of $M(p)$ represents the number of these machines that are idle at the current global state.

In a net graph as shown in Fig. 1.1, a marking can be represented by a distribution of *tokens* in all places, i.e., putting $M(p)$ tokens (those black dots in Fig. 1.1) into the place p.

A marking is sometimes denoted as a $|P|$-dimensional nonnegative integer vector indexed by P and every element represents the number of tokens in the corresponding place at this marking,[1] e.g. the marking $M = [1, 0, 6, 0]$ indexed by $P = \{p_1, \ p_2, \ p_3, \ p_4\}$ represents that the places p_1, p_2, p_3 and p_4 have 1, 0, 6, and 0 tokens at M, respectively. For convenience, a marking M is sometimes denoted as a multi-set:

$$M = \{M(p) \cdot p \mid \forall p \in P\} \ \text{ or } \ M = \sum_{p \in P} M(p) \cdot p.$$

For the above example, it may be written as the multi-set $M = \{p_1, 6p_3\} = p_1 + 6p_3$.

Given a subset P' of P and a marking M over P, $M(P')$ denotes the number of tokens in P', i.e.,

$$M(P') = \sum_{p \in P'} M(p).$$

Definition 1.3 (*Elementary Net System*) A net $N = (P, T, F)$ with an *initial marking* M_0 is called an *elementary net system* and denoted as $(N, M_0) \triangleq (P, T, F, M_0)$.

Notice that all contents in this book are based on elementary net systems; but for convenience, we call an elementary net system as a *Petri net* in this book although the definition of Petri net in the Petri net theory is broader than elementary net systems.

Example 1.2 Figure 1.1 illustrates a Petri net whose initial marking is

$$\{p_{j,1}, \ r_j \mid j \in \mathbb{N}_5^+\}$$

which means such a concrete global state of the system: each philosopher is thinking and each chopstick is idle.

In this book, when we say a state of a Petri net or a model, this state means a concrete global state.

1.1.2   Firable Transition Sequence and Interleaving Semantics

Through the flow relation of a Petri net, the condition enabling every transition and the result after firing it can be specified. Therefore, at a given marking, it is decidable whether a transition is enabled or not, and firing an enabled transition leads to a new

---

[1] It is assumed that there is a total order on P so that the j-th entry in a vector corresponds to the j-th place in this ordered set.

marking. The changes of these states illustrate all possible runs and thus can be used to check if some runs result in some errors. An error means an incorrect design of the related system, and thus finding errors is one of important targets of formally modelling and analysing a system. Next, the formal definitions of enabling and firing a transition are shown.

1.1.2.1   Pre-set and Post-set

A transition t is called an *input transition* of a place p and p is called an *output place* of t if $(t, p) \in F$.

Input place and *output transition* can be defined similarly.

Given a net $N = (P, T, F)$ and a node $x \in P \cup T$, the *pre-set* $^\bullet x$ and *post-set* $x^\bullet$ of x are defined, respectively, as follows:

$$^\bullet x \triangleq \{y \in P \cup T \mid (y, x) \in F\},$$

$$x^\bullet \triangleq \{y \in P \cup T \mid (x, y) \in F\}.$$

These concepts can be extended to a set $X \subseteq P \cup T$, i.e., the *pre-set* and *post-set* of X are defined, respectively, as follows:

$$^\bullet X \triangleq \bigcup_{x \in X} {}^\bullet x,$$

$$X^\bullet \triangleq \bigcup_{x \in X} x^\bullet.$$

For a transition, its pre-set and post-set represent the pre-conditions and post-conditions of executing it (if it is viewed as an event). For a place, its pre-set and post-set stand for those producers and consumers that produce or consume a token into or out of this place (if it is viewed as a warehouse storing goods).

1.1.2.2   Enabling and Firing Rules

A place $p \in P$ is *marked* at a marking M if

$$M(p) > 0.$$

A transition $t \in T$ is *enabled* at M if every input place of it is marked at M, i.e.,

$$\forall p \in {}^\bullet t : M(p) > 0,$$

which is denoted by the following notation:

$$M[t\rangle.$$

On the contrary, a transition $t \in T$ is *disabled* at M if there exists an input place of it is not marked at M, i.e.,

$$\exists p \in {}^\bullet t : M(p) = 0,$$

which is denoted by the following notation:

$$\neg M[t\rangle.$$

Firing an enabled transition t at M leads to a new marking M' such that for each place $p \in P$:

$$M'(p) = \begin{cases} M(p) - 1 & p \in {}^\bullet t \setminus t^\bullet \\ M(p) + 1 & p \in t^\bullet \setminus {}^\bullet t \\ M(p) & otherwise \end{cases},$$

which is denoted by the following notation:

$$M[t\rangle M'.$$

Example 1.3 At the initial marking

$$M_0 = \{p_{j,1}, \, r_j \mid j \in \mathbb{N}_5^+\}$$

of the Petri net in Fig. 1.1, the transition $t_{j,1}$ is enabled for each $j \in \mathbb{N}_5^+$ since each place in ${}^\bullet t_{j,1} = \{p_{j,1}, r_j\}$ is marked at M_0. Firing the transition $t_{5,1}$ at the initial marking leads to the marking

$$\{p_{5,2}, \, p_{j,1}, r_j \mid j \in \mathbb{N}_4^+\}$$

since it deletes two tokens out of places $p_{5,1}$ and r_5 and produces a token into the place $p_{5,2}$. At this new marking, transitions $t_{1,1}$ and $t_{5,2}$ are both enabled, but firing each of them disables another one since the token in the common place r_1 is deleted.

1.1.2.3 Firable Transition Sequence

Based on these rules, all *firable transition sequences* starting from a given marking M and being of a given step length k ($k \in \mathbb{N}^+$) can be constructed, which is due to the finiteness of both the set of places and transitions and the given length k.[2] A transition sequence $\sigma = t_1 t_2 \ldots t_k$ is *firable* at M if there are markings $M_1, \ldots,$ and M_k such that

$$M[t_1\rangle M_1[t_2\rangle \cdots \rangle M_{k-1}[t_k\rangle M_k.$$

This evolution sequence can be written simply as follows:

$$M[\sigma\rangle M_k.$$

The step length of a firable transition sequence is the number of transitions in it. Generally, a firable transition sequence of a Petri net means a sequential run from the initial marking of the Petri net.

Example 1.4 At the initial marking

$$M_0 = \{p_{j,1}, \ r_j \mid j \in \mathbb{N}_5^+\}$$

of the Petri net in Fig. 1.1, the following transition sequence is firable and makes the system return its initial marking again:

$$t_{5,1} t_{5,2} t_{5,3}.$$

The following two transition sequences are both firable at the initial marking and lead to the same marking:

$$t_{5,1} t_{1,1} t_{1,2} t_{1,3} \ \text{and} \ t_{1,1} t_{5,1} t_{1,2} t_{1,3}.$$

As illustrated in the above example, transitions $t_{1,1}$ and $t_{5,1}$ are both enabled at the initial marking, and firing each of them does not disable another one. In other words, the two transitions can be fired concurrently. However, the above method of representing the dynamic behaviours of a Petri net, i.e., *firable transition sequences*, does not characterise the concurrence of the two transitions since a firable transition sequence is strictly sequential. A sequence $\cdots a \cdots b \cdots$ means that the occurrence of a is earlier than the occurrence of b. Consequently, two transitions, say a and b, lead to two firable sequences $\cdots a \cdots b \cdots$ and $\cdots b \cdots a \cdots$ in some arbitrary order if they can be fired concurrently at a marking. This kind of representation of dynamic behaviours of a concurrent system is called *interleaving semantics*, which obviously cannot represent the concurrence of actions.

---

[2] A firable transition sequence is called a *sequential run* in [1].

1.1.3 Process and Concurrency Semantics

The concurrency semantics of a Petri net is explained on the basis of a so-called *causal net*.

For representing the concurrence in the dynamic behaviours of a Petri net, the concept *process* was proposed on the basis of *causal net* [1].

> When we discuss the concurrency semantics, we assume that a Petri net has no transition without any input place.

Such a transition that has no input place can be fired infinitely and does not consume any token. If a Petri net has such a transition, then a new place can be added into the Petri net such that this place and this transition are connected by a *self-loop* and a token is put into the place, which can guarantee that this new Petri net and the original one have the same behaviours as shown in Fig. 1.2. Therefore, this assumption is reasonable and that will brings convenience for the definition of concurrency semantics.

1.1.3.1 Causal Net

Definition 1.4 (*Causal Net*) A net $N = (P, T, F)$ is a *causal net* if the following conditions hold:

1. for each $p \in P$ we have $|{}^\bullet p| \leq 1$ and $|p^\bullet| \leq 1$, and
2. F^+ is irreflexive where F^+ is the transitive closure of the flow relation F.

Example 1.5 Figure 1.3 illustrates a causal net where for the label of each node, the part outside the bracket is the name of the node and the part inside the bracket needs to be neglected temporally.

A causal net has no choice/branch according to $|p^\bullet| \leq 1$ for each $p \in P$, i.e., in every local state there does not exist two transitions in a conflict relation. Later, we will introduce *conflict*. The result caused by every transition in a causal net is independent due to $|{}^\bullet p| \leq 1$ for each $p \in P$.

The second constraint in Definition 1.4 means that a causal net has no loop. Isolated places such as $c_{2,1}$ in Fig. 1.3 can exist in a causal net, but any isolated transition cannot exist according to the previous assumption that the preset of each transition is not empty.

(a)   (b)

Fig. 1.2 Equal transitions: **a** no input; and **b** self-loop

Fig. 1.3 A causal net that is a *process* of the Petri net in Fig. 1.1. For the label on each node, the part outside the bracket is the name of the node, and the part inside the bracket is the name of the corresponding node in Fig. 1.1

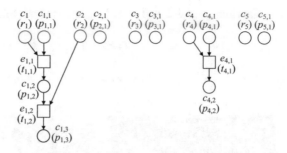

Given a causal net N, its *outset* and *end* are defined as follows:

$$°N \triangleq \{x \in P \cup T \mid {}^{\bullet}x = \emptyset\},$$

$$N° \triangleq \{x \in P \cup T \mid x^{\bullet} = \emptyset\}.$$

In fact, the definition of outset is equivalent to the following one

$$°N = \{x \in P \mid {}^{\bullet}x = \emptyset\},$$

since the preset of every transition is not empty according to the previous assumption.

Example 1.6 The outset and end of the causal net in Fig. 1.3 are

- $°N = \{c_1, c_{1,1}, c_2, c_{2,1}, c_3, c_{3,1}, c_4, c_{4,1}, c_5, c_{5,1}\}$,
- $N° = \{c_{1,3}, c_{2,1}, c_3, c_{3,1}, c_{4,2}, c_5, c_{5,1}\}$.

1.1.3.2 Causality and Concurrency Relations

Since a causal net has neither branch nor loop, any two different transitions in it are either causal (F^+) or concurrent. Next, more general definitions about these relations are shown.

Given a causal net $N = (P, T, F)$, two binary relations $\mathbf{li} \subseteq (P \cup T) \times (P \cup T)$ and $\mathbf{co} \subseteq (P \cup T) \times (P \cup T)$ of N are defined as follows:

$$\mathbf{li} \triangleq \{(x, y) \mid (x, y) \in F^+ \vee (y, x) \in F^+ \vee x = y\},$$

$$\mathbf{co} \triangleq \{(x, y) \mid (x, y) \notin F^+ \wedge (y, x) \notin F^+)\}.$$

$\mathbf{li}$ denotes all node pairs in the causality relation (i.e., $F^+ \cup (F^+)^{-1}$) as well as all node pairs (x, x) where $x \in P \cup T$. $\mathbf{co}$ denotes all node pairs in the concurrency relation as well as all node pairs (x, x) where $x \in P \cup T$.

Obviously, both **li** and **co** are symmetric and reflexive, and we have:

$$\mathbf{li} \cap \mathbf{co} = \{(x, x) \mid x \in P \cup T\},$$

$$\mathbf{li} \cup \mathbf{co} = (P \cup T) \times (P \cup T).$$

For any two different nodes in a causal net, they are either in the causality relation (**li**) or in the concurrency relation (**co**). Although it is not necessarily reasonable to consider that a node and itself are in the both causality and concurrency relations, it will be convenient to define *line* and *cut* whose special case contains only one node. For convenience, **li** and **co** are called as *causality relation* and *concurrency relation*, respectively.

Definition 1.5 (*Line, Cut and Slice of Causal Net*) Let $N = (P, T, F)$ be a causal net, $X \subseteq P \cup T$ be a node set, and **li** and **co** be the causality and concurrency relations, respectively.

1. X is called a *line* of N if for any two nodes x and y in X we have $(x, y) \in \mathbf{li}$ and for each $z \in (P \cup T) \setminus X$ there exists $x \in X$ such that $(x, z) \notin \mathbf{li}$ and $(z, x) \notin \mathbf{li}$.
2. X is called a *cut* of N if for any two nodes x and y in X we have $(x, y) \in \mathbf{co}$ and for each $z \in (P \cup T) \setminus X$ there exists $x \in X$ such that $(x, z) \notin \mathbf{co}$ and $(z, x) \notin \mathbf{co}$.
3. X is called a *slice* of N if X is cut and $X \subseteq P$.

Example 1.7 The causal net in Fig. 1.3 has 10 lines as follows:

- $\{c_1, e_{1,1}, c_{1,2}, e_{1,2}, c_{1,3}\}$,
- $\{c_{1,1}, e_{1,1}, c_{1,2}, e_{1,2}, c_{1,3}\}$,
- $\{c_2, e_{1,2}, c_{1,3}\}$,
- $\{c_4, e_{4,1}, c_{4,2}\}$,
- $\{c_{4,1}, e_{4,1}, c_{4,2}\}$,
- $\{c_{2,1}\}$,
- $\{c_3\}$,
- $\{c_{3,1}\}$,
- $\{c_5\}$,
- $\{c_{5,1}\}$.

Example 1.8 The causal net in Fig. 1.3 has 15 cuts as follows:

- $\{c_1, c_{1,1}, c_2, c_{2,1}, c_3, c_{3,1}, c_4, c_{4,1}, c_5, c_{5,1}\}$,
- $\{c_1, c_{1,1}, c_2, c_{2,1}, c_3, c_{3,1}, c_{4,2}, c_5, c_{5,1}\}$,
- $\{c_1, c_{1,1}, c_2, c_{2,1}, c_3, c_{3,1}, e_{4,1}, c_5, c_{5,1}\}$,
- $\{c_{1,2}, c_2, c_{2,1}, c_3, c_{3,1}, c_4, c_{4,1}, c_5, c_{5,1}\}$,

- $\{c_{1,2},\ c_2,\ c_{2,1},\ c_3,\ c_{3,1},\ c_{4,2},\ c_5,\ c_{5,1}\}$,
- $\{c_{1,2},\ c_2,\ c_{2,1},\ c_3,\ c_{3,1},\ e_{4,1},\ c_5,\ c_{5,1}\}$,
- $\{c_{1,3},\ c_{2,1},\ c_3,\ c_{3,1},\ c_4,\ c_{4,1},\ c_5,\ c_{5,1}\}$,
- $\{c_{1,3},\ c_{2,1},\ c_3,\ c_{3,1},\ c_{4,2},\ c_5,\ c_{5,1}\}$,
- $\{c_{1,3},\ c_{2,1},\ c_3,\ c_{3,1},\ e_{4,1},\ c_5,\ c_{5,1}\}$,
- $\{e_{1,1},\ c_2,\ c_{2,1},\ c_3,\ c_{3,1},\ c_4,\ c_{4,1},\ c_5,\ c_{5,1}\}$,
- $\{e_{1,1},\ c_2,\ c_{2,1},\ c_3,\ c_{3,1},\ c_{4,2},\ c_5,\ c_{5,1}\}$,
- $\{e_{1,1},\ c_2,\ c_{2,1},\ c_3,\ c_{3,1},\ e_{4,1},\ c_5,\ c_{5,1}\}$,
- $\{e_{1,2},\ c_{2,1},\ c_3,\ c_{3,1},\ c_4,\ c_{4,1},\ c_5,\ c_{5,1}\}$,
- $\{e_{1,2},\ c_{2,1},\ c_3,\ c_{3,1},\ c_{4,2},\ c_5,\ c_{5,1}\}$,
- $\{e_{1,2},\ c_{2,1},\ c_3,\ c_{3,1},\ e_{4,1},\ c_5,\ c_{5,1}\}$.

Example 1.9 The causal net in Fig. 1.3 has 6 slices as follows:

- $\{c_1,\ c_{1,1},\ c_2,\ c_{2,1},\ c_3,\ c_{3,1},\ c_4,\ c_{4,1},\ c_5,\ c_{5,1}\}$,
- $\{c_1,\ c_{1,1},\ c_2,\ c_{2,1},\ c_3,\ c_{3,1},\ c_{4,2},\ c_5,\ c_{5,1}\}$,
- $\{c_{1,2},\ c_2,\ c_{2,1},\ c_3,\ c_{3,1},\ c_4,\ c_{4,1},\ c_5,\ c_{5,1}\}$,
- $\{c_{1,2},\ c_2,\ c_{2,1},\ c_3,\ c_{3,1},\ c_{4,2},\ c_5,\ c_{5,1}\}$,
- $\{c_{1,3},\ c_{2,1},\ c_3,\ c_{3,1},\ c_4,\ c_{4,1},\ c_5,\ c_{5,1}\}$,
- $\{c_{1,3},\ c_{2,1},\ c_3,\ c_{3,1},\ c_{4,2},\ c_5,\ c_{5,1}\}$.

A line (cut, respectively) is a maximal set of pairwise nodes in the causality (concurrency, respectively) relation, and a slice is such a maximal set of pairwise places in the concurrency relation. In the definition of line, the constraint

$$\forall z \in (P \cup T) \setminus X, \exists x \in X : (x, z) \notin \mathbf{li} \wedge (z, x) \notin \mathbf{li}$$

can be simplified into

$$\forall z \in (P \cup T) \setminus X, \exists x \in X : (x, z) \notin \mathbf{li}$$

since $(x, z) \notin \mathbf{li}$ implies $(z, x) \notin \mathbf{li}$ according to the symmetry of $\mathbf{li}$, and that is similar to the definition of cut. Additionally, $^{\circ}N$ is obviously a slice, but N° is a cut but not necessarily a slice if a transition without output is in it.

1.1.3.3 Definition of Process

Based on causal nets, the concurrency semantics of Petri nets can be defined that represents the concurrent behaviours. In other words, every run of a Petri net containing the concurrence of transitions is represented by a causal net.

For convenience, a Petri net is denoted as $(N, M_0) = (P_N, T_N, F_N, M_0)$ and a causal net is denoted as $K = (P_K, T_K, F_K)$ in this section.

Definition 1.6 (*Process*) Given a Petri net $(N, M_0) = (P_N, T_N, F_N, M_0)$ and a causal net $K = (P_K, T_K, F_K)$ with $(P_N \cup T_N) \cap (P_K \cup T_K) = \emptyset$, (K, ρ) is called a *process* of (N, M_0) if there is a mapping $\rho: K \to N$ satisfying the following conditions:

1. for each $p \in P_K$ we have $\rho(p) \in P_N$,
2. for each $t \in T_K$ we have $\rho(t) \in T_N$, $\rho({}^\bullet t) = {}^\bullet \rho(t)$, and $\rho(t^\bullet) = \rho(t)^\bullet$,
3. $\rho({}^\circ K) = M_0$.

Example 1.10 The causal net in Fig. 1.3 illustrates a *process* of the Petri net in Fig. 1.1, and the labels of all nodes shows the mapping ρ where the part outside a bracket is the name of a node of the causal net and the part inside the bracket is the name of the corresponding node of the Petri net.

A *process* is called a *distributed run* in [1].

> A *process*, as a kind of *concurrency semantics* of Petri nets, represents a run of a Petri net, containing all possible concurrences in the run.

> For a firable transition sequence of a Petri net, there exists a *process* of the Petri net corresponding to it because a causal net is easily constructed according to the initial marking and the transition sequence. But, a *process* of a Petri net possibly corresponds to multiple firable transition sequences because the concurrent transitions in the *process* corresponds to multiple interleaving transition sequences. Each slice of a *process* corresponds to a marking of the related Petri net.

Example 1.11 The *process* in Fig. 1.3 corresponds to three firable transition sequences as follows:

$$t_{4,1}t_{1,1}t_{1,2}, \quad t_{1,1}t_{4,1}t_{1,2}, \quad t_{1,1}t_{1,2}t_{4,1}.$$

It has 6 slices as shown in Example 1.9 that exactly correspond to 6 markings yielded by firing the three transition sequences.

In Definition 1.6, $\rho({}^\circ K)$ is a multi-set. In other words, when a place p of a Petri net has multiple tokens at the initial marking, ${}^\circ K$ has multiple places corresponding to p.

Example 1.12 Figure 1.4a shows a Petri net where place r_1 has two tokens at the initial marking, and (b) shows one *process* of it. For this *process*, we have

Fig. 1.4 a A Petri net in which place r_1 has two tokens in the initial marking. **b** A *process* of the Petri net in **a** in which places c_2 and c_3 both correspond to r_1

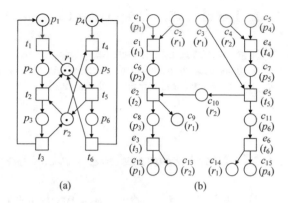

(a)   (b)

$$^\circ K = \{c_1,\ c_2,\ c_3,\ c_4,\ c_5\}$$

and

$$\rho(^\circ K) = \{p_1,\ p_4,\ 2r_1,\ r_2\}$$

due to

$$\rho(c_2) = \rho(c_3) = r_1.$$

From this *process*, we can see that (e_1, e_4), (e_1, e_5), (e_1, e_6), (e_2, e_6), and (e_3, e_6) are all in the concurrency relation. In the original Petri net, those transition pairs corresponding to them can also be fired concurrently at the related markings.

When a Petri net has some transitions that can be executed repeatedly, it has infinitely many *processes*.

If the outset and the end of a *process* correspond to the same marking, the *process* is called a *scenario* [1], which means that the behaviours represented by the scenario can be executed repeatedly.

Example 1.13 The *process* in Fig. 1.4b is a scenario since its outset and end correspond to the same marking, but the *process* in Fig. 1.3 is not a scenario because its outset and end do not correspond to the same marking.

In what follows, we introduce two methods usually used in analysing the correctness of the behaviours of Petri nets. The two methods are based on the interleaving and concurrency semantics, respectively.

1.2   Reachability Graph and Coverability Graph

This section introduces reachability graphs and coverability graphs of Petri nets as well as their generation algorithms. At the same time, reachability, coverability, boundedness, and unboundedness are introduced. More details can be found in [2, 7].

1.2.1   Reachability and Reachability Graph

The reachability graph method is based on the interleaving semantics. In other words, the reachability graph of a Petri net is used to represent all possible states (markings) and firable transition sequences of the Petri net. It has a root node corresponding to the initial marking of the Petri net, and each directed path corresponds to a firable transition sequence. Before describing the definition of reachability graph, the following concept is first introduced.

> A marking M_k is *reachable* from a marking M in a Petri net if either $M_k = M$ or there is a firable transition sequence σ such that $M[\sigma\rangle M_k$.

The set of all markings reachable from a marking M in a net N is denoted as $R(N, M)$. Consequently, given a Petri net (N, M_0), $R(N, M_0)$ denotes all markings reachable from the initial marking M_0.

Definition 1.7 (*Reachability Graph*) Given a Petri net $(N, M_0) = (P, T, F, M_0)$, its *reachability graph*, denoted as $RG(N, M_0) \triangleq (R(N, M_0), E)$, is a labeled digraph satisfying

1. $R(N, M_0)$ is all reachable markings of (N, M_0) that forms the node set of the reachability graph where M_0 is called the *root node*,[3] and
2. $E \triangleq \{(M, M')_t \mid M, M' \in R(N, M_0), t \in T: M[t\rangle M'\}$ is the set of all directed edges that are labeled by transitions.

Example 1.14 Figure 1.5 shows the reachability graph of the Petri net in Fig. 1.4a. Through this graph, it can be seen that there are two interleaving transition sequences $t_1 t_4$ and $t_4 t_1$ starting from the initial marking, while t_1 and t_4 can be concurrently fired at the initial marking.

There are possibly multiple directed edges from a node to another one in a reachability graph but they are labeled by different transitions. This case is generally caused

---

[3] In fact, we should have defined the set of node names and markings should have been labelled to these nodes. Because there is a one-to-one correspondence between node names and markings, we directly use markings as the node names here.

Fig. 1.5 The reachability
graph of the Petri net in
Fig. 1.4a

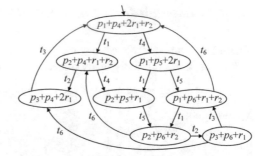

Algorithm 1.1: Producing the Reachability Graph of a Petri Net.

Input: A Petri net $(N, M_0) = (P, T, F, M_0)$;
Output: The reachability graph $RG(N, M_0) = (R(N, M_0), E)$;
1 $Q_{new} \leftarrow \emptyset$;
2 $R(N, M_0) \leftarrow \emptyset$;
3 $E \leftarrow \emptyset$;
4 **PutTail**(Q_{new}, M_0);
5 **while** $Q_{new} \neq \emptyset$ **do**
6 $\quad$ $M \leftarrow$ **GetHead**(Q_{new});
7 $\quad$ $R(N, M_0) \leftarrow R(N, M_0) \cup \{M\}$;
8 $\quad$ **for** *each* $t \in T$ *such that* $M[t\rangle$ **do**
9 $\quad\quad$ Produce marking M' such that $M[t\rangle M'$;
10 $\quad\quad$ **if** $M' \notin Q_{new} \cup R(N, M_0)$ **then**
11 $\quad\quad\quad$ **PutTail**(Q_{new}, M');
12 $\quad\quad$ $E \leftarrow E \cup \{(M, M')_t\}$;

---

by those transitions with self-loop. For example, if two transitions, say a and b, are added into the Petri net in Fig. 1.4 that, respectively, are connected with places p_2 and p_6, then the above case can occur in the related reachability graph. That is the reason why a labeled directed edge is represented by $(M, M')_t$ in Definition 1.7.

Algorithm 1.1 describes how to construct the reachability graph of a Petri net [7]. The idea is to utilise a queue to store those new markings. The queue is initialised by M_0 (Line 4). A marking is pulled out of the queue (Line 6), and each succeeding marking (successor) of it is produced (Line 9). Only when a succeeding marking is new (Line 10), it is put into the queue (Line 11). Whenever a succeeding marking is new or not, a labeled directed edge from the original marking to the succeeding one is produced (Line 12). In this algorithm, the function **GetHead** is to pull the head element out of a queue, and the function **PutTail** is to insert an element into the tail of a queue.

1.2.2   Unboundedness and Coverability Graph

When the number of tokens in some places of a Petri net can be increased infinitely in its runs, it has infinitely many reachable markings and thus is *unbounded*.

A Petri net $(N, M_0) = (P, T, F, M_0)$ is *k-bounded* for a given integer $k \in \mathbb{N}^+$ if the following holds:

$$\forall p \in P, \forall M \in R(N, M_0) : M(p) \leq k.$$

A Petri net is *bounded* if it is k-bounded for some k; and otherwise *unbounded*. In other words, boundedness means that the set $R(N, M_0)$ is finite, and unboundedness means $R(N, M_0)$ is infinite. A Petri net is *safe* if it is 1-bounded.

Example 1.15 Figure 1.1 shows a safe Petri net, Fig. 1.4a shows a 2-bounded but not safe Petri net since place r_1 can contain up to two tokens and each of others has at most one token, and Fig. 1.6a, b show two unbounded Petri nets since the place p_2 can contain as many tokens as we want.

The reachability graph of an unbounded Petri net has infinitely many nodes, and thus Algorithm 1.1 cannot be terminated for an unbounded Petri net, bringing troubles to the analysis of such a Petri net.

The *coverability graph* of an unbounded Petri net can be constructed based on its reachability graph, containing finitely many nodes. The idea is that when a new marking covers an old one that occurs in the directed path from the initial marking to this new one, then this new marking is modified into another form with an *infinite number* ∞ that satisfies the following rule:

$$\forall n \in \mathbb{N}: n < \infty \land \pm n + \infty = \infty \pm n = \infty.$$

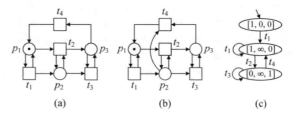

(a)   (b)   (c)

Fig. 1.6   a A live Petri net; **b** a non-live Petri net; and **c** their coverability graph. Obviously, it is not easy to analyse liveness via the coverability graph method. This example is from [2]

Algorithm 1.2: Producing the Coverability Graph of an Unbounded Petri Net.

Input: A Petri net $(N, M_0) = (P, T, F, M_0)$;
Output: The coverability graph $CG(N, M_0) = (R(N, M_0)_\infty, E)$;
1 $Q_{new} \leftarrow \emptyset$;
2 $R(N, M_0)_\infty \leftarrow \emptyset$;
3 $E \leftarrow \emptyset$;
4 **PutTail**(Q_{new}, M_0);
5 **while** $Q_{new} \neq \emptyset$ **do**
6 | $M \leftarrow$ **GetHead**(Q_{new});
7 | $R(N, M_0)_\infty \leftarrow R(N, M_0)_\infty \cup \{M\}$;
8 | **for** *each $t \in T$ such that $M[t\rangle$* **do**
9 | | Produce marking M' such that $M[t\rangle M'$;
10 | | **if** $M' \notin Q_{new} \cup R(N, M_0)_\infty$ **then**
11 | | | **if** *M' covers a marking M'' occurring in the path from M_0 to M'* **then**
12 | | | $M' \leftarrow M' \uplus_\infty M''$;
13 | | **PutTail**(Q_{new}, M');
14 | $E \leftarrow E \cup \{(M, M')_t\}$;

For instance, if a new marking $[1, 1, 2, 0]$ covers an old one $[1, 0, 1, 0]$, then this new one is modified into the form $[1, \infty, \infty, 0]$. Here, a marking M' (with ∞) *covers* another one M (with ∞) if and only if the following holds:

$$\forall p : M'(p) \geq M(p) \wedge \exists p : M'(p) > M(p).$$

When a marking M' (with ∞) covers another one M'' (with ∞), the operation

$$M' \uplus_\infty M''$$

leads to a new marking M''' (with ∞) such that for each place p:

$$M'''(p) = \begin{cases} M'(p) & M'(p) = M''(p) \\ \infty & M'(p) > M''(p) \end{cases}.$$

Algorithm 1.2 describes the procedure of producing the coverability graph of a Petri net [7]. It can always be terminated, i.e., the coverability graph of an unbounded Petri net is finite, due to the finiteness of the number of places.

Example 1.16 Figure 1.6a, b show two different Petri nets that are both unbounded. The first one is live, but the second one is not. Liveness will be introduced later. However, their coverability graphs are the same as shown in Fig. 1.6c, which means that the coverability graph of an unbounded Petri net cannot necessarily represent its dynamic behaviours accurately.

For a bounded Petri net, its reachability graph is exactly a finite-state machine [8] (if we do not consider the final states of a finite-state machine). And given a finite-state machine, a bounded Petri net is easily constructed corresponding to it where each state is viewed as a place, and for each transition of the machine a net transition is constructed to connect the related two places [7]. Therefore, bounded Petri nets and finite-state machines can be viewed as a pair of identical concepts (when considering the sequential behaviours) [7].

> Even for bounded Petri nets, the reachability graph method generally face the *state explosion problem*, i.e., the state space of a system grows exponentially or much worse with the size of the system, bringing much challenge to their analysis.

For instance, if there are n concurrent transitions in a marking, then 2^n different succeeding markings can be produced according to the interleaving semantics. In fact, these markings and all possible runs can be represented by a special net (similar to a causal net) with $k \cdot n$ nodes according to the concurrent semantics, where $k \cdot n$ is a linear function of n. Next, such a concurrent-semantics-based method is introduced.

1.3 Unfolding

This section introduces the unfolding method of Petri net, and presents an algorithm of producing a finite complete prefix of the unfolding of a bounded Petri net as well as an algorithm of producing a finite (but not complete) prefix of the unfolding of an unbounded Petri net. More details of the unfolding method, readers may read the literatures [3, 9].

1.3.1 Branching Process

A *process* is defined on the basis of a causal net, while a *branching process* is defined in view of a so-called *occurrence net*.

1.3.1.1 Occurrence Net and Conflict Relation

First, the concept of causal net is extended to *occurrence net* that permits *conflict* since the constraint of $|p^\bullet| \leq 1$ is deleted.

Definition 1.8 (*Occurrence Net*) A net $N = (P, T, F)$ is an *occurrence net* if the following conditions hold:

1. for each $p \in P$ we have $|{}^\bullet p| \leq 1$,
2. F^+ is irreflexive, and

Fig. 1.7 A branching process of the Petri net in Fig. 1.1. In the branching process, there a branch/choice at the place c_j for each $j \in \mathbb{N}_5^+$

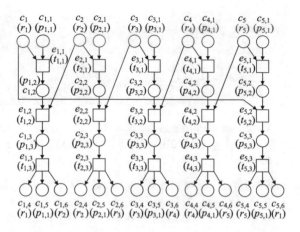

3. no transition is *self-conflict*, i.e., for each transition $t \in T$, there do not exist two distinct transitions t_1 and t_2 such that

$$(t_1, t) \in F^+ \wedge (t_2, t) \in F^+ \wedge {}^\bullet t_1 \cap {}^\bullet t_2 \neq \emptyset.$$

Two nodes x and y of an occurrence net $N = (P, T, F)$ are *conflict* if there are two distinct transitions t_1 and t_2 such that

$$(t_1, x) \in F^+ \wedge (t_2, y) \in F^+ \wedge {}^\bullet t_1 \cap {}^\bullet t_2 \neq \emptyset,$$

i.e., there exist two branches starting from the same place and respectively reaching the two nodes. Obviously, if there is one token in such a place, then choosing and executing one of the two branches will deny the chance of executing anther one. Consequently, if there were a self-conflict transition in an occurrence net, the transition could never be fired, which is the reason why the definition of occurrence net needs the third condition.

Example 1.17 Figure 1.7 shows an occurrence net in which the place c_j has more than one output for each $j \in \mathbb{N}_5^+$. These pairs like

$$(e_{4,2}, e_{5,1}), (e_{5,1}, e_{4,2}), (e_{4,2}, c_{5,2}) (c_{4,3}, c_{5,2})$$

are all in a conflict relation since the two nodes in each pair, respectively, come from the two different branches of the place c_5.

Different from the definition of concurrency relation **co** of a causal net, the concurrency relation of an occurrence net needs to exclude not only the causality relation but also the conflict relation. For convenience, we use the same notation to represent them.

Given an occurrence net $N = (P, T, F)$, three binary relations $\mathbf{li} \subseteq (P \cup T) \times (P \cup T)$, $\mathbf{cf} \subseteq (P \cup T) \times (P \cup T)$ and $\mathbf{co} \subseteq (P \cup T) \times (P \cup T)$ of N are defined as follows:

- $\mathbf{li} \triangleq \{(x, y) \mid (x, y) \in F^+ \vee (y, x) \in F^+ \vee x = y\}$,
- $\mathbf{cf} \triangleq \{(x, y) \mid \exists t_1, t_2 \in T : t_1 \neq t_2 \wedge (t_1, x) \in F^+ \wedge (t_2, y) \in F^+ \wedge {}^\bullet t_1 \cap {}^\bullet t_2 \neq \emptyset\}$,
- $\mathbf{co} \triangleq \{(x, y) \mid ((x, y) \notin \mathbf{li} \wedge (x, y) \notin \mathbf{cf}) \vee x = y\}$.

Just like **li** and **co** of a causal net, **li** and **co** of an occurrence net also contain all pairs (x, x); but **cf** of an occurrence net does not contain these reflexive pairs. Obviously, **cf** is symmetric.

Definition 1.9 (*Line, Cut and Slice of Occurrence Net*) Let $N = (P, T, F)$ be an occurrence net and $X \subseteq P \cup T$ be a node set. **li**, **cf**, and **co** are the causality, conflict, and concurrency relations of N, respectively.

1. X is called a *line* of N if for any two nodes x and y in X we have $(x, y) \in \mathbf{li}$, and for each $z \in (P \cup T) \setminus X$ there exists $x \in X$ such that $(x, z) \notin \mathbf{li}$ and $(z, x) \notin \mathbf{li}$.
2. X is called a *cut* of N if for any two nodes x and y in X we have $(x, y) \in \mathbf{co}$ and for each $z \in (P \cup T) \setminus X$ there exists $x \in X$ such that $(x, z) \notin \mathbf{co}$ and $(z, x) \notin \mathbf{co}$.
3. X is called a *slice* of N if X is cut and $X \subseteq P$.

Any nonempty subset of a slice is generally called a *co-set*.

1.3.1.2   Definition of Branching Process

For convenience, a Petri net and an occurrence net are denoted by $(N, M_0) = (P_N, T_N, F_N, M_0)$ and $O = (P_O, T_O, F_O)$ in this section, respectively.

Definition 1.10 (*Branching Process*) Given a Petri net $(N, M_0) = (P_N, T_N, F_N, M_0)$ and an occurrence net $O = (P_O, T_O, F_O)$ with $(P_N \cup T_N) \cap (P_O \cup T_O) = \emptyset$, (O, ρ) is called a *branching process* of (N, M_0) if there is a mapping $\rho : O \to N$ satisfying the following conditions:

1. for each $p \in P_O$ we have $\rho(p) \in P_N$,
2. for each $t \in T_O$ we have $\rho(t) \in T_N$, $\rho({}^\bullet t) = {}^\bullet \rho(t)$, and $\rho(t^\bullet) = \rho(t)^\bullet$,

Fig. 1.8 a A Petri net; and **b** an occurrence net that is not a branching process of the Petri net in **a** because the left branch and the right one represent the same *process*

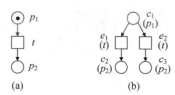

3. $\rho(^\circ O) = M_0$ where $^\circ O = \{p \in P_O \mid {}^\bullet p = \emptyset\}$,
4. for any two transitions t_1 and t_2 in T_O, if ${}^\bullet t_1 = {}^\bullet t_2$ and $\rho(t_1) = \rho(t_2)$ then $t_1 = t_2$.

Example 1.18 Figure 1.7 shows a branching process of the Petri net in Fig. 1.1. In Fig. 1.8, the occurrent net in (b) is not a branching process of the Petri net in (a) although it satisfies the first three conditions of Definition 1.10; however, it does not satisfy the fourth one, i.e., the two branches mean the same *process*.

Obviously, a *process* of a Petri net is also a branching process in which there is no other branch, while the fourth constraint of Definition 1.10 is to avoid that the same *process* is represented (as different branches) multiple times in a branching process.

Theorem 1.1 ([10]) *Let* $(O, \rho) = (P_O, T_O, F_O, \rho)$ *be a branching process of Petri net* $(N, M_0) = (P_N, T_N, F_N, M_0)$, *and* $X \subseteq P_O$ *is a slice of* O. *Then, there exists a reachable marking* $M \in R(N, M_0)$ *such that*

$$\rho(X) = M.$$

Theorem 1.2 ([10]) *Let* $(N, M_0) = (P_N, T_N, F_N, M_0)$ *be a Petri net and* $M \in R(N, M_0)$ *be a reachable marking. Then, there exists a branching process* $(O, \rho) = (P_O, T_O, F_O, \rho)$ *of* (N, M_0) *and a slice* $X \subseteq P_O$ *such that*

$$\rho(X) = M.$$

In fact, the above conclusions provided in the literature [10] are about *process*, while here they are about branching process. Just as mentioned above, a branching process represents a set of *processes* and the same *process* is not represented multiple times in the branching process. Thus, the above conclusions hold still. These conclusions are also easily understood via the algorithm of producing branching processes that will be introduced in what follows.

1.3.1.3 Prefix and Canonical Coding of Branching Process

Definition 1.11 (*Prefix*) Given two branching processes $(O_j, \rho_j) = (P_{O_j}, T_{O_j}, F_{O_j}, \rho_j)$ of a Petri net $(N, M_0) = (P_N, T_N, F_N, M_0)$ where $j \in \mathbb{N}_2^+$, the branching process (O_1, ρ_1) is a *prefix* of the branching process (O_2, ρ_2) if the following holds:

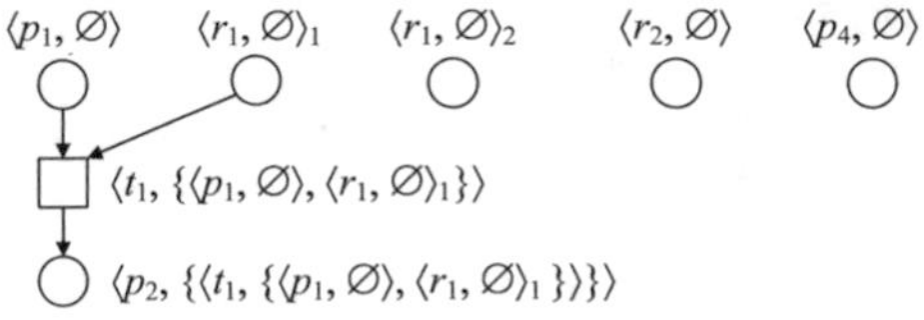

Fig. 1.9 A branching process of the Petri net in Fig. 1.4a, which is coded by the canonical coding method

$\langle p_1, \varnothing \rangle$ $\langle r_1, \varnothing \rangle_1$ $\langle r_1, \varnothing \rangle_2$ $\langle r_2, \varnothing \rangle$ $\langle p_4, \varnothing \rangle$

$\langle t_1, \{\langle p_1, \varnothing \rangle, \langle r_1, \varnothing \rangle_1\}\rangle$

$\langle p_2, \{\langle t_1, \{\langle p_1, \varnothing \rangle, \langle r_1, \varnothing \rangle_1\}\rangle\}\rangle$

$$P_{O_1} \subseteq P_{O_2} \wedge T_{O_1} \subseteq T_{O_2}.$$

For instance, the branching process in Fig. 1.3 is a prefix of the one in Fig. 1.7. Note that infinitely many isomorphic occurrence nets may represent the same branching process. In order to represent all isomorphic branching processes in a uniform way, a so-called *canonical coding* method can be used to label all transitions and places of a branching process, i.e.,

$$\forall x \in P_O \cup T_O : cod(x) \triangleq \langle \rho(x), \{cod(y) \mid y \in {}^\bullet x\}\rangle.$$

For instance, Fig. 1.9 shows a branching process in which nodes are coded by this canonical coding method. The code of a node has shown the corresponding relation from the node to the original Petri net as well as the inputs of the node. For each element $x \in {}^\circ O$, its code is $\langle \rho(x), \varnothing \rangle$ since ${}^\bullet x = \varnothing$.

> If a place p of a Petri net has k ($k > 1$) tokens in the initial marking, then there will be k places in ${}^\circ O$ corresponding to it and their codes can be distinct via different subscripts, e.g. $\langle p, \varnothing \rangle_1, \cdots,$ and $\langle p, \varnothing \rangle_k$ as shown in Fig. 1.9.

Obviously, a code becomes very long with the increase of the depth of a branching process, and thus for simplicity, we still use the label form as shown in Figs. 1.3 and 1.7; but the canonical codings bring convenience for describing some definitions and algorithms that will be seen soon.

Although the definition of prefix only requires $P_{O_1} \subseteq P_{O_2}$ and $T_{O_1} \subseteq T_{O_2}$, it actually implies the following facts [11]:

- $(t \in T_{O_1} \wedge ((p, t) \in F_{O_2} \vee (t, p) \in F_{O_2})) \Rightarrow p \in P_{O_1},$[4]
- $(p \in P_{O_1} \wedge (t, p) \in F_{O_1}) \Rightarrow t \in T_{O_2}$, and
- ρ_1 is a restriction of ρ_2 onto $P_{O_1} \cup T_{O_1}$, i.e.,

$$\forall x \in P_{O_1} \cup T_{O_1} : \rho_1(x) = \rho_2(x).$$

---

[4] The symbol $\Rightarrow$ is the Boolean operator of *implication*. $a \Rightarrow b$ holds if and only if $\neg a \vee b$ holds.

1.3.1.4   Unfolding and Possible Extension

In fact, the set of all branching processes of a Petri net is a *partially ordered set* w.r.t. the binary relation *prefix*, i.e., it is reflexive, antisymmetric, and transitive. Furthermore, it forms a *complete lattice*. Therefore, it has a unique greatest element that is called the *unfolding* of the Petri net.

Algorithm 1.3: Producing the Unfolding of a Petri Net.

Input: A Petri net $(N, M_0) = (P_N, T_N, F_N, M_0)$ where $M_0 = \{p_1, \ldots, p_n\}$;
Output: The unfolding $(O, \rho) = (P_O, T_O, F_O, \rho)$ of (N, M_0);
1 $P_O \leftarrow \{\langle p_1, \emptyset \rangle, \cdots, \langle p_n, \emptyset \rangle\}$;
2 $T_O \leftarrow \emptyset$;
3 $F_O \leftarrow \emptyset$;
4 Compute all possible extensions PE of the current branching process;
5 **while** $PE \neq \emptyset$ **do**
6 | Choose a possible extension $\langle t, X \rangle$ from PE;
7 | Compute all outputs P of $\langle t, X \rangle$, i.e., $P \leftarrow \{\langle p, \{\langle t, X \rangle\} \rangle \mid p \in P_N \wedge p \in t^\bullet\}$;
8 | $P_O \leftarrow P_O \cup P$;
9 | $T_O \leftarrow T_O \cup \{\langle t, X \rangle\}$;
10 | $F_O \leftarrow F_O \cup \{(x, \langle t, X \rangle) \mid x \in X\} \cup \{(\langle t, X \rangle, x) \mid x \in P\}$;
11 | Compute all possible extensions PE of the current branching process;

Given a Petri net, its unfolding can be produced by Algorithm 1.3 [11]. The idea of this algorithm is to repeat adding a *possible extension* to a given branching process and thus a new branching process is generated. Note that the initial branching process is very simple since it only corresponds to the initial marking of the given Petri net. The possible extension of a branching process is defined as follows:

Definition 1.12 (*Possible Extension*) Let (O, ρ) be a branching process of Petri net (N, M_0). Then, $\langle t, X \rangle$ is a *possible extension* of (O, ρ) if the following conditions hold:

1. X is a co-set of O, t is a transition of N and $\rho(X) = {}^\bullet t$, and
2. $\langle t, X \rangle$ has not occurred in (O, ρ).

Example 1.19 Figure 1.9 has two possible extensions:

- $\langle t_4, \{\langle p_4, \emptyset \rangle, \langle r_2, \emptyset \rangle\} \rangle$,
- $\langle t_2, \{\langle p_2, \{\langle t_1, \{\langle p_1, \emptyset \rangle, \langle r_1, \emptyset \rangle_1\} \rangle\} \rangle\}, \langle r_2, \emptyset \rangle\} \rangle$.

Although

$$\langle t_1, \{\langle p_1, \emptyset \rangle, \langle r_1, \emptyset \rangle_2\} \rangle$$

is also a possible extension of this branching process, any *process* starting from it is equivalent to the one starting from

$$\langle t_1, \{\langle p_1, \emptyset \rangle, \langle r_1, \emptyset \rangle_1\}\rangle,$$

and thus this possible extension can be omitted from the aspect of the analysis technique.

In Algorithm 1.3, when a possible extension is added, its output places and the related arcs are also produced.

> When a Petri net is unbounded, its unfolding is infinite and thus Algorithm 1.3 cannot be terminated. Even for a bounded Petri net, its unfolding is still infinite if it has an infinite firable transition sequence.

Therefore, a finite prefix of an unfolding is necessary if this method is used to analyze a system, which is introduced in what follows.

1.3.2 Finite Complete Prefix

For a bounded Petri net, a finite complete prefix of its unfolding means that all reachable markings and all transitions of the Petri net can be found in the finite complete prefix. Algorithm 1.4 describes such a method of producing a finite complete prefix of a bounded Petri net, and the idea is that an infinite *process* is actually represented by a minimal scenario. In other words, all succeeding transitions of a special transition are deleted from an infinite *process* if the marking resulted in by the occurrence of this special transition has existed before its occurrence. Such a transition, called a *cut-off transition* of the unfolding, is still remained in a finite complete prefix but all succeeding nodes of it (except its output places) are cut off.

Given a transition t of an unfolding $(O, \rho) = (P_O, T_O, F_O, \rho)$, the *process ended at t* is defined as

$$\lceil t \rfloor \triangleq \{t' \in T_O \mid (t', t) \in \mathbf{li}\}.$$

In [12], $\lceil t \rfloor$ is called a *local configuration* of t which is represented by the notation $[t]$ there. But, to distinguish it from the notation $[N]$ representing an incidence matrix in the next chapter, we use the notation $\lceil t \rfloor$ here.

Although $\lceil t \rfloor$ only represents a set of transitions, they actually correspond to a *process*. The marking reached by such a *process* is denoted as

$$Mark(\lceil t \rfloor) \triangleq \rho((^\circ O \cup \lceil t \rfloor^\bullet) \setminus {}^\bullet \lceil t \rfloor).$$

Algorithm 1.4: Producing a Finite Complete Prefix of a Bounded Petri Net.

Input: A bounded Petri net $(N, M_0) = (P_N, T_N, F_N, M_0)$ where $M_0 = \{p_1, \ldots, p_n\}$;
Output: A finite complete prefix $(O, \rho) = (P_O, T_O, F_O, \rho)$ of (N, M_0);

1   $P_O \leftarrow \{\langle p_1, \emptyset \rangle, \cdots, \langle p_n, \emptyset \rangle\}$;
2   $T_O \leftarrow \emptyset$;
3   $F_O \leftarrow \emptyset$;
4   $Cutoff \leftarrow \emptyset$;
5   Compute all possible extensions PE of the current branching process;
6   **while** $PE \neq \emptyset$ **do**
7   Choose a possible extension $\langle t, X \rangle$ from PE;
8   **if** $\langle t, X \rangle$ *is a cut-off transition of the current branching process* **then**
9   **if** $\langle t, X \rangle \notin Cutoff$ **then**
10            $Cutoff \leftarrow Cutoff \cup \{\langle t, X \rangle\}$;
11         $PE \leftarrow PE \setminus \{\langle t, X \rangle\}$;
12   **else**
13   Compute all outputs P of $\langle t, X \rangle$, i.e., $P \leftarrow \{\langle p, \{\langle t, X \rangle\} \rangle \mid p \in P_N \wedge p \in t^\bullet\}$;
14         $P_O \leftarrow P_O \cup P$;
15         $T_O \leftarrow T_O \cup \{\langle t, X \rangle\}$;
16         $F_O \leftarrow F_O \cup \{(x, \langle t, X \rangle) \mid x \in X\} \cup \{(\langle t, X \rangle, x) \mid x \in P\}$;
17   Compute all possible extensions PE of the current branching process;

18   **for** *each* $\langle t, X \rangle \in Cutoff$ **do**
19      $P \leftarrow \{\langle p, \{\langle t, X \rangle\} \rangle \mid p \in P_N \wedge p \in t^\bullet\}$;
20      $P_O \leftarrow P_O \cup P$;
21      $T_O \leftarrow T_O \cup \{\langle t, X \rangle\}$;
22      $F_O \leftarrow F_O \cup \{(x, \langle t, X \rangle) \mid x \in X\} \cup \{(\langle t, X \rangle, x) \mid x \in P\}$;

Example 1.20 In Fig. 1.7, $\lceil e_{1,3} \rfloor = \{e_{1,1}, e_{1,2}, e_{1,3}\}$ and thus $\lceil e_{1,3} \rfloor$, $\lceil e_{1,3} \rfloor^\bullet$, $^\circ O$ and the related arcs form a *process*. And we have

$$Mark(\lceil e_{1,3} \rfloor) = \rho(\{c_{1,4}, c_{1,5}, c_{1,6}, c_{2,1}, c_3, c_{3,1}, c_4, c_{4,1}, c_5, c_{5,1}\}) = \{r_j, \ p_{j,1} \mid j \in \mathbb{N}_5^+\}.$$

For a bounded Petri net, a transition t of its unfolding is called a *cut-off transition* of the unfolding if the following holds:

$$(\exists t' \in \lceil t \rfloor : t' \neq t \wedge Mark(\lceil t' \rfloor) = Mark(\lceil t \rfloor)) \vee Mark(\lceil t \rfloor) = M_0.$$

Example 1.21 In Fig. 1.7, $e_{j,3}$ is a cut-off transition since

$$Mark(\lceil e_{j,3} \rfloor) = M_0$$

where $j \in \mathbb{N}_5^+$.

Fig. 1.10 A finite complete prefix of the unfolding of the Petri net in Fig. 1.4a. Transitions e_3, e_6 and e_8 are cut-off transitions and added into the prefix at last according to Algorithm 1.4

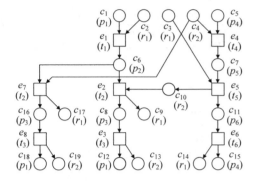

Example 1.22 The branching process in Fig. 1.7 is exactly a finite complete prefix of the unfolding of the Petri net in Fig. 1.1. The branching process in Fig. 1.10 is exactly a finite complete prefix of the unfolding of the Petri net in Fig. 1.4a.

1.3.3 Finite Prefix

For an unbounded Petri net, a finite complete prefix cannot be provided, because any finite prefix only represents finite states but an unbounded Petri net has infinitely many states. Here, a finite prefix can be provided to represent finite behaviours of an unbounded Petri net. Algorithm 1.4 can still be utilised to produce such a finite prefix, but the only difference lies in the definition of cut-off transition.

> For an unbounded Petri net, a transition t of its unfolding is called a *cut-off transition* of the unfolding if the following holds:
>
> $$\exists t' \in \lceil t \rfloor : t' \neq t \wedge Mark(\lceil t' \rfloor) \leq Mark(\lceil t \rfloor).$$

This definition means that the marking resulted in by such a transition covers another one prior to it. This definition does not consider the case of covering the initial marking, which can make more markings and transitions occur in a finite prefix.

Example 1.23 Figure 1.11a, b show two finite prefixes corresponding to the unbounded Petri nets in Fig. 1.6a, b, respectively. If the definition of cut-off transition also required $Mark(\lceil t \rfloor) \geq M_0$, the two finite prefixes would have only one transition $e_1(t_1)$ and thus other transitions would not occur in them.

The idea of the above algorithms of producing finite (complete) prefixes is simple, i.e., an infinite *process* is cut at the position where the slice covers an earlier one;

Fig. 1.11 **a** A finite prefix of the unfolding of the unbounded Petri net in Fig. 1.6a, in which e_2, e_4 and e_7 are the cut-off transitions; and **b** finite prefix of the unfolding of the unbounded Petri net in Fig. 1.6b, in which e_2, e_5 and e_6 are the cut-off transitions

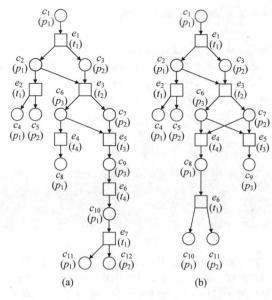

(a)   (b)

Fig. 1.12 **a** A Petri net; **b** the finite complete prefix produced by Algorithm 1.4 for the Petri net in **a**; and **c** a small-size finite complete prefix of the Petri net in **a**

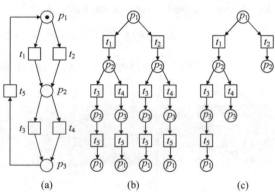

(a)   (b)   (c)

but they maybe result in a large-size finite prefix since many similar *processes* are possibly kept in it. For instance, Fig. 1.12b is the finite complete prefix produced by Algorithm 1.4 for the Petri net in Fig. 1.12a, but some branches are unnecessary and thus can be cut as shown in Fig. 1.12c.

1.4 Basic Properties and Computation Complexity

This section introduces some basic properties of Petri nets including reachability, coverability, boundedness, liveness, deadlock, and persistence, as well as the complexity of determining them.

1.4.1 Some Basic Properties

In the above sections, reachability, coverability and boundedness have been introduced. Now, deadlock, liveness, and persistence are introduced.

A reachable marking M of a Petri net (N, M_0) is called a *deadlock* if it disables any transition, i.e., the following holds:

$$M \in R(N, M_0) \land \forall t \in T : \neg M[t\rangle.$$

A deadlock of a system means such a situation in which no action can be executed.

Example 1.24 For the Petri net in Fig. 1.1, the reachable marking $\{p_{j,2} \mid j \in \mathbb{N}_5^+\}$ is a deadlock, which means that each philosopher holds the left chopstick but waits for the right one. For the Petri net in Fig. 1.6b, the marking $\{p_3\}$ reached by firing the transition sequence $t_1 t_2 t_3$ at the initial marking $\{p_1\}$ is a deadlock, too.

A Petri net $(N, M_0) = (P, T, F, M_0)$ is *live* if the following holds:

$$\forall t \in T, \forall M \in R(N, M_0), \exists M' \in R(N, M) : M'[t\rangle.$$

Liveness means that each action has always a chance to be executed at any state, and thus implies deadlock-free.

Example 1.25 The Petri net in Fig. 1.6a is live, but the Petri nets in Figs. 1.1 and 1.6b are not.

A Petri net $(N, M_0) = (P, T, F, M_0)$ is *persistent* if the following holds:

$$\forall t_1, t_2 \in T, \forall M \in R(N, M_0) : (M[t_1\rangle \land M[t_2\rangle \land t_1 \neq t_2) \Rightarrow M[t_1 t_2\rangle.$$

Persistence means that for any two actions that both can be executed at a state, executing one does not disable the other.

Example 1.26 The Petri net in Fig. 1.4a is persistent if its initial marking is $\{p_1, p_4, 2r_1, 2r_2\}$.

For a bounded Petri net, the problems of deciding reachability, liveness and deadlock can all be solved through its reachability graph or unfolding; however, the complexity is generally very high because of the so-called state space explosion problem. In what follows some complexity problems in the Petri net theory are reviewed briefly.

1.4.2 Computation Complexity of Deciding Basic Properties

Through the definitions of coverability, boundedness, liveness, deadlock and persistence, it can be seen that they are closely related to reachability, and the problem of determining if a Petri net satisfies a given property can be reducible to the problem of determining if a related reachability holds [7, 13]. Therefore, reachability is generally thought of as the most basic property in the Petri net theory.

It has been proven that the reachability problem for Petri nets is decidable [14], but the complexity is very high. Here, the *reachability problem* for Petri nets is to ask *given a Petri net and a marking, if the marking is reachable in the Petri net?*

Theorem 1.3 ([15, 16]) *The reachability problem for Petri nets is not elementary, but there is a TOWER lower bound and an ACKERMANN upper bound.*

An accurate result of the complexity of the reachability problem is still open.[5] Similarly, the accurate computation complexity of determining coverability, liveness, deadlock and persistence for Petri nets are also open; but a lower bound, EXPSPACE-hard, has been proven [17, 18]. Since the liveness and coverability problems are equivalent to the reachability problem [7], the lower bounds of their complexity are also TOWER based on Theorem 1.3. For the safe subclass, a precise bound can usually be provided for these problems.

By constructing the reachability graph of a given Petri net, one can decide whether the Petri net is bounded or not. If a Petri net is bounded, then it has a finite number of markings and thus Algorithm 1.1 can be terminated normally. If a Petri net is unbounded, then there always exists a reachable marking that properly covers one of its ancestors and thus the execution of the algorithm can be terminated in advance. However, the complexity of the boundedness problem for Petri nets is still very high. Here, the boundedness problem for Petri nets is to ask: *given a Petri net, if it is bounded?* It has been proven that this problem is also EXPSPACE-hard [17, 19]. But, the *k-boundedness problem* for Petri nets is PSPACE-complete [17] where this problem is to ask: *given a Petri net and a positive integer k, if the Petri net is k-bounded?* Obviously, the complexity of the two problems is different: when finding the number of tokens in some place in a reachable marking is greater than k, one can decide that the Petri net is not k-bounded but cannot decide if it is bounded or not.

---

[5] Recently, Czerwiński and Orlikowski proved that the problem is ACKERMANN-complete, which can be downloaded in https://arxiv.org/abs/2104.13866 but has not been published in any peer-reviewed journal or conference until the publication of this book.

For the safe subclass of Petri nets, the following problems are all PSPACE-complete
[17, 18]:

- The *reachability problem* for safe Petri nets: *given a safe Petri net and a marking,*
  if the marking is reachable in the Petri net?
- The *coverability problem* for safe Petri nets: *given a safe Petri net and a marking*
  M, if there is a reachable marking M' of the Petri net such that M' ≥ M?
- The *liveness problem* for safe Petri nets: *given a safe Petri net, if it is live?*
- The *deadlock problem* for safe Petri nets: *given a safe Petri net, if it has a deadlock?*
- The *persistence problem* for safe Petri nets: *given a safe Petri net, if it is persistent?*

1.5 Application

The dining philosophers problem describes such a class of problems: a group of
processes (e.g. assembly processes in a manufacturing system or threads in an oper-
ating system) require finitely many resources and finally release them, but deadlocks
possibly occur when these resources are allocated improperly.[6] An improper allocation can result in a *circular waits*: each of these processes holds some resources
but is waiting for some other ones held by other processes. As shown in Fig. 1.1,
each philosopher holds its left chopstick but is waiting for the right one. In such a
situation, every process cannot do anything but wait.

Fig. 1.13 A deadlock-free
model of the Dining
Philosophers Problem in
which each philosopher can
eat only when s/he holds the
two chopsticks at one
time [1, 2]

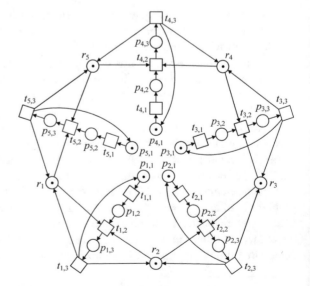

---

[6] The word "process" here is not the "process" in Sects. 1.1 and 1.3. To distinguish them, they are
italic in Sects. 1.1 and 1.3.

There is such an allocation policy that can avoid deadlocks: all resources which a process needs are allocated to it at one time.

For example, the two chopsticks a philosopher needs are allocated to this philosopher when s/he is ready for eating and they are available [1, 2]. The Petri net in Fig. 1.13 illustrates this resource allocation policy in which place $p_{j,2}$ can be viewed as the *ready-for-eating* state and firing the transition $t_{j,2}$ means that the related two chopsticks are allocated to the j-th philosopher at one time. It can be verified through its reachability graph or unfolding that this model has no deadlock and is live, and thus this policy does work.

1.6 Summary and Further Reading

This chapter introduces the basic knowledge of elementary net systems, including two kinds of semantics as well as the related analysis ways. The formal definitions of some basic properties including deadlock, liveness, reachability, (un)boundedness, coverability, and persistence are provided.

The basic concepts and interleaving semantics (sequential run) of Petri nets can be found in [1]. The reachability graph, the coverability graph and the algorithms of producing them can be founded in [7].

Process (distributed run), causal net, and the related concepts such as concurrency, line, cut, slice, and scenario can be found in [1, 10]. Branching process, occurrence net, unfolding, possible extension and the algorithm of producing unfolding can be found in [11, 12].

Many studies have shown that the size and shapes of a finite complete prefix of a safe Petri net are related to a pre-defined *order* of local configurations. A good order can decrease the size of a finite prefix. For more related results, readers may read literatures [3, 11, 12, 20, 21].

For unbounded Petri nets, we also proposed another algorithm of producing their finite prefixes and used them to decide the soundness of workflow nets [6, 22]. A workflow net is a special Petri net in which there are two special places called *source place* without input and *sink place* without output, and each node is in some path from the source place to the sink place. This subclass was defined by Dr. Wil van der Aalst [23] and its soundness means that a system is deadlock-free and livelock-free. It will later be introduced in detail in this book.

Recently, we studied the complexity of the soundness problem for many kinds of subclasses of workflow nets as well as the complexity of the deadlock problem for some subclasses of Petri nets, and readers interested in them can read literatures [24–30].

References

1. Reisig, W.: *Understanding Petri Nets: Modeling Techniques, Analysis Methods, Case Studies.* Springer-Verlag, Berlin Heidelberg (2013)
2. Wu, Z.H.: *Introduction to Petri Nets.* China Machine Press, Beijing (2006) (in Chinese)
3. Esparza, J., Heljanko, K.: *Unfoldings: A Partial-Order Approach to Model Checking.* Springer-Verlag, Berlin Heidelberg (2008)
4. Dijkstra, E.W.: Hierarchical ordering of sequential processes. *Acta Informatica* 1: 115–138 (1972)
5. Hoare, C.A.R.: *Communicating Sequential Processes.* Prentice Hall, Englewood Cliffs (2015)
6. Liu, G.J., Reisig, W., Jiang, C.J., Zhou, M.C.: A branching-process-based method to check soundness of workflow systems. *IEEE Access* 4: 4104–4118 (2016)
7. Peterson, J.L.: *Petri Net Theory and the Modeling of Systems.* Prentice-Hall International, Englewood Cliffs (1980)
8. Cassandras C.G., Lafortune, S.: *Introduction to Discrete Event Systems* (Third Edition). Springer Nature Switzerland AG (2021)
9. Liu, G.J.: *Primary Unfoldings of Petri Nets: A Model Checking Method for Concurrent Systems.* China Science Press, Beijing (2020) (in Chinese)
10. Goltz, U., Reisig, W.: The non-sequential behaviour of petri nets. *Information and Control* 57: 125–147 (1983)
11. Engelfriet, J.: Branching Processes of Petri nets. *Acta Informatica* 28: 575–591 (1991)
12. Esparza, J., Römer, S., Vogler, W.: An improvement of McMillan's unfolding algorithm. *Formal Methods in System Design* 20: 285–310 (2002)
13. Jones, N.D., Landweber, L.H., Lien, Y.E.: Complexity of some problems in Petri nets. *Theoretical Computer Science* 4: 277–299 (1976)
14. Karp, R.M., Miller, R.E.: Parallel program schemata. *Journal of Computer and System Sciences* 3: 147–195 (1969)
15. Czerwiński, W., Lasota, S., Lazić, R., Laroux, J., Mazowiecki, F.: The reachability problem for Petri nets is not elementary. In: *the 51st Annual ACM SIGACT Symposium on Theory of Computing*, pp. 24–33 (2019)
16. Leroux, J., Schmitz, S.: Reachability in Vector Addition Systems Is Primitive-recursive in Fixed Dimension. In: *the 34th Annual ACM/IEEE Symposium on Logic in Computer Science*, pp. 1–13 (2019)
17. Lipton, R.J.: The reachability problem requires exponential space. *Technical Report* 62, Yale University (1976)
18. Cheng, A., Esparza, J., Palsberg, J.: Complexity results for 1-safe nets. *Theoretical Computer Science* 147: 117–136 (1995)
19. Yen, H.C.: A unified approach for deciding the existence of certain petri net paths. *Information and Computation* 96: 119–137 (1992)
20. McMillan, K.L.: Using unfolding to avoid the state explosion problem in the verification of asynchronous circuits. In: *the 4th Workshop on Computer Aided Verification*, pp. 164–174 (1993)
21. McMillan, K.L.: A technique of sate space search based on unfolding. *Formal methods in System Design* 6: 45–65 (1995)
22. Liu, G., Zhang, K., Jiang, C.J.: Deciding the deadlock and livelock in a Petri net with a target marking based on Its basic unfolding. In: *the 16the IEEE International Conference on Algorithms and Architectures for Parallel Processing*, pp. 98–105 (2016)
23. van der Aalst, W.: Structural characterisations of sound workflow nets. *Computing Science Report* 96/23, Eindhoven University of Technology (1996)
24. Liu, G.J., Sun, J., Liu, Y., Dong, J.S.: Complexity of the soundness problem of bounded workflow nets. In: *the 33rd International Conference on Application and Theory of Petri Nets and Concurrency*, pp. 92–107 (2012)

25. Liu, G.J., Jiang, C.J., Zhou, M.C., Ohta, A.: The liveness of WS^{3}PR: complexity and decision. *IEICE Transactions on Fundamentals of Electronics, Communications and Computer Sciences* E96-A: 1783–1793 (2013)
26. Liu, G.J.: Some complexity results for the soundness problem of workflow nets. *IEEE Transactions on Services Computing* 7: 322–328 (2014)
27. Liu, G.J., Sun, J., Liu, Y., Dong, J.S.: Complexity of the soundness problem of workflow nets. *Fundamenta Informaticae* 131: 81–101 (2014)
28. Liu, G.J., Jiang, C.J.: Co-NP-hardness of the soundness problem for asymmetric-choice workflow nets. *IEEE Transactions on Systems, Man and Cybernetics: Systems* 45: 1201–1204 (2015)
29. Liu, G.J.: Complexity of the deadlock problem for Petri nets modelling resource allocation systems. *Information Sciences* 363: 190–197 (2016)
30. Liu, G.J.: PSPACE-completeness of the soundness problem of safe asymmetric-choice workflow nets. In: *the 41st International Conference on Application and Theory of Petri Nets and Concurrency*, pp. 196–216 (2020)

Chapter 2
Structural Characteristics of Petri Nets

2.1 Incidence Matrix and State Equation

This section introduces incidence matrix and state equations that can also be used to represent the dynamic sequential behaviours of a Petri net. For more details, one can read [1].

2.1.1 Incidence Matrix

Since a net can be viewed as a directed bipartite graph, its structure can be represented by a matrix according to the graph theory. In such a matrix, columns and rows correspond to two different categories of nodes, respectively.

Definition 2.1 (*Incidence Matrix*) The *incidence matrix* of a net $N = (P, T, F)$ is a $|T| \times |P|$ matrix $[N]$ indexed by T and P such that

$$[N](t, p) = \begin{cases} 1 & (t, p) \in F \wedge (p, t) \notin F \\ -1 & (t, p) \notin F \wedge (p, t) \in F \ . \\ 0 & \text{otherwise} \end{cases}$$

Here, rows and columns of an incidence matrix correspond to transitions and places, respectively. An order of all transitions (respectively all places) is provided in advance.

© The Author(s), under exclusive license to Springer Nature Singapore Pte Ltd. 2022   35
G. Liu, *Petri Nets*,
https://doi.org/10.1007/978-981-19-6309-4_2

Example 2.1 The incidence matrix of the net in Fig. 1.4 is

$$
\begin{array}{c}
\\
t_1 \\
t_2 \\
t_3 \\
t_4 \\
t_5 \\
t_6
\end{array}
\begin{array}{c}
\begin{array}{cccccccc} p_1 & p_2 & p_3 & p_4 & p_5 & p_6 & r_1 & r_2 \end{array} \\
\left[
\begin{array}{cccccccc}
-1 & 1 & 0 & 0 & 0 & 0 & -1 & 0 \\
0 & -1 & 1 & 0 & 0 & 0 & 1 & -1 \\
1 & 0 & -1 & 0 & 0 & 0 & 0 & 1 \\
0 & 0 & 0 & -1 & 1 & 0 & 0 & -1 \\
0 & 0 & 0 & 0 & -1 & 1 & -1 & 1 \\
0 & 0 & 0 & 1 & 0 & -1 & 1 & 0
\end{array}
\right]
\end{array}
$$

where the kth row corresponds to t_k $(k \in \mathbb{N}_6^+)$, the jth column corresponds to p_j $(j \in \mathbb{N}_6^+)$, and the 7th and 8th columns correspond to r_1 and r_2, respectively.

When a net has no self-loop, its structure can be represented by its incidence matrix. However, when a net has a self-loop, its incidence matrix cannot represent it.

Example 2.2 For the net in Fig. 1.6a, its incidence matrix is

$$
\begin{array}{c}
\\
t_1 \\
t_2 \\
t_3 \\
t_4
\end{array}
\begin{array}{c}
\begin{array}{ccc} p_1 & p_2 & p_3 \end{array} \\
\left[
\begin{array}{ccc}
0 & 1 & 0 \\
-1 & 0 & 1 \\
0 & -1 & 0 \\
1 & 0 & -1
\end{array}
\right]
\end{array}
$$

in which an element 0 cannot represent if there is a self-loop or not between the related transition and place.

To solve the problem, the structure of a net can be represented by two matrices $[N]^+$ and $[N]^-$ defined as follows:

$$
[N]^+(t, p) = \begin{cases} 1 & (t, p) \in F \\ 0 & \text{otherwise} \end{cases},
$$

$$
[N]^-(t, p) = \begin{cases} 1 & (p, t) \in F \\ 0 & \text{otherwise} \end{cases}.
$$

$[N]^+$ is called the *input matrix* of N, representing the input relation from transitions to places; and $[N]^-$ is called the *output matrix* of N, representing the output relation from places to transitions.

Example 2.3 For the net in Fig. 1.6a, its $[N]^+$ and $[N]^-$, respectively, are shown as follows:

$$
\begin{array}{c}
\quad\quad p_1 \quad p_2 \quad p_3 \\
\begin{array}{c} t_1 \\ t_2 \\ t_3 \\ t_4 \end{array}
\left[
\begin{array}{ccc}
1 & 1 & 0 \\
0 & 1 & 1 \\
0 & 0 & 1 \\
1 & 0 & 0
\end{array}
\right]
\end{array}
$$

and

$$
\begin{array}{c}
\quad\quad p_1 \quad p_2 \quad p_3 \\
\begin{array}{c} t_1 \\ t_2 \\ t_3 \\ t_4 \end{array}
\left[
\begin{array}{ccc}
1 & 0 & 0 \\
1 & 1 & 0 \\
0 & 1 & 1 \\
0 & 0 & 1
\end{array}
\right]
\end{array}
.
$$

There is the following relation among incidence matrix, input matrix and output matrix of a net:

$$[N] = [N]^+ - [N]^-.$$

2.1.2 State Equation

Obviously, given a net N and a marking M of N, a transition t of N is enabled at M if and only if

$$M \geq [N]^-(t)$$

where M is viewed as a vector over the place set and $[N]^-(t)$ is the row vector corresponding to t in $[N]^-$.

The new marking M' yielded by firing an enabled transition t at M can be computed by the following formula:

$$M' = M - [N]^-(t) + [N]^+(t) = M + [N](t)$$

where $[N]^-(t)$ represents the tokens removed from the input places of t and $[N]^+(t)$ represents the tokens produced into the output places of t.

The above shows a new definition form of enabling and firing a transition, based on the input and output matrices of a net.

As a result, reachability should also be able to be reflected by some calculations over an incidence matrix.

After firing a firable transition sequence tt' at M, the reached marking can be calculated by the following formula:

$$M' = M + [N](t) + [N](t') = M + [0, \ldots, 0, \overset{t}{1}, 0, \ldots, 0, \overset{t'}{1}, 0, \ldots, 0] \cdot [N].$$

Generally speaking, the marking reached by firing a firable transition sequence σ at M can be calculated by the following formula:

$$M' = M + \psi(\sigma) \cdot [N]$$

where $\psi(\sigma)$ is the *firing count vector* of σ indexed by the transition set T. The firing count vector of a firable transition sequence is a vector, indexed by the transition set, such that every element corresponds to the number of times that the related transition occurs in this sequence.

Example 2.4 In the Petri net in Fig. 1.6a,

$$t_1 t_1 t_1 t_2 t_3 t_3$$

is a firable transition sequence at the initial marking $[1, 0, 0]$, and the yielded marking is $[0, 1, 1]$ after firing this sequence. The firing count vector of this sequence is

$$[3, 1, 2, 0].$$

Obviously, the following formula holds:

$$[0, 1, 1] = [1, 0, 0] + [3, 1, 2, 0] \cdot \begin{bmatrix} 0 & 1 & 0 \\ -1 & 0 & 1 \\ 0 & -1 & 0 \\ 1 & 0 & -1 \end{bmatrix}.$$

From the above example we can see that given a Petri net (N, M_0) and a marking M, if $M \in R(N, M_0)$, then there exists one solution of nonnegative integers for the following so-called *state equation*:

$$M = M_0 + X \cdot [N].$$

This is because the firing count vector corresponding to any firable transition sequence from M_0 to M is such a solution.

Fig. 2.1   a A simple Petri
net; **b** the incidence matrix of
the net in **a**; **c** the dual net of
the net in **a** where for the
label of each node, the part
outside the bracket is the
name of the node and the part
inside the bracket is the name
of the corresponding node
in **a**; and **d** the incidence
matrix of the net in **c**.

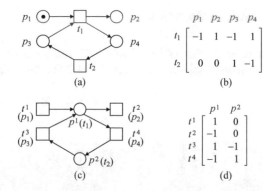

If a marking is reachable from another one, then the related state equation has
a nonnegative integer solution; but, when a state equation has a nonnegative
integer solution, this does not necessarily means that a marking is reachable
from another one.

Example 2.5   For the Petri net in Fig. 2.1a, the marking $[0, 1, 0, 0] = \{p_2\}$ is not
reachable, but the state equation

$$[0, 1, 0, 0] = [1, 0, 0, 0] + X \cdot [N]$$

has a nonnegative integer solution $[1, 1]$. The reason is that there are some loop
structures so that a place without token first lends a transition a *token* and then *it* is
given back by another transition in the loop, e.g. the loop $p_3 t_1 p_4 t_2 p_3$ in Fig. 2.1a.

Sometimes, such a loop structure corresponds to a *T-invariant* or *repetitive vector*
such that the original marking is covered after the related transitions are fired. On
the other hand, there are also some subclasses of Petri nets such that a marking is
reachable if and only if its state equation has a nonnegative solution. In what follows,
we will introduce these related contents.

2.2   Invariant

This section first introduces the concepts of T-invariant and P-invariant as well as
the dynamic behaviour features behind them. At last, a famous method of computing
them is illustrated that is through the row reduction of incidence matrices. More
details can be found in [1–3].

2.2.1   T-Invariant

Definition 2.2 (*T-invariant*) A $|T|$-dimensional nonnegative integer vector X is a *T-invariant* of a net $N = (P, T, F)$ if the following holds:

$$X \neq \bar{0} \wedge X \cdot [N] = \bar{0}.$$

The first $\bar{0}$ in Definition 2.2 is a $|T|$-dimensional vector where every entry is 0, i.e.,

$$\bar{0}_{1 \times |T|} \triangleq [0, \cdots, 0]_{1 \times |T|};$$

and the second $\bar{0}$ is $|P|$-dimensional, i.e.,

$$\bar{0}_{1 \times |P|} \triangleq [0, \ldots, 0]_{1 \times |P|};$$

but they are both abbreviated as $\bar{0}$ when no ambiguity is introduced.

Example 2.6 In the net in Fig. 1.6a, $[1,\ 1,\ 1,\ 1]$ is a T-invariant because of

$$[1,\ 1,\ 1,\ 1] \cdot \begin{bmatrix} 0 & 1 & 0 \\ -1 & 0 & 1 \\ 0 & -1 & 0 \\ 1 & 0 & -1 \end{bmatrix} = \bar{0},$$

and in the net in Fig. 1.6b, $[2,\ 1,\ 1,\ 1]$ is a T-invariant because of

$$[2,\ 1,\ 1,\ 1] \cdot \begin{bmatrix} 0 & 1 & 0 \\ -1 & 0 & 1 \\ 0 & -1 & 0 \\ 1 & -1 & -1 \end{bmatrix} = \bar{0}.$$

Given a T-invariant, if any less vector is not a T-invariant, then it is called a *minimal T-invariant*.

According to Definition 2.2 it can easily be known that any multiple of a T-invariant is still a T-invariant, and the sum of any two T-invariants is also a T-invariant. Since a T-invariant is limited in the field of nonnegative integers, a net has finitely many minimal T-invariants but infinitely many T-invariants (if it has T-invariants).

A finite set of T-invariants of a net is called a *basis* of all T-invariants of the net if any T-invariant of the net can be linearly represented by them with nonnegative rational coefficients. All minimal T-invariants of a net form a *basis* of all T-invariants.

Theorem 2.1 ([1]) *Given a net, any T-invariant of it can be represented by a linear combination of all minimal T-invariants of it with nonnegative rational coefficients.*

This proof is left to readers interested in it.

The *support* of a T-invariant X is defined as

$$||X|| \triangleq \{t \in T \mid X(t) > 0\}.$$

Example 2.7 The net in Fig. 1.6a has exactly two minimal T-invariants

$$[1, \; 0, \; 1, \; 0] \text{ and } [0, \; 1, \; 0, \; 1]$$

whose supports are

$$\{t_1, \; t_3\} \text{ and } \{t_2, \; t_4\},$$

respectively. It is easy to verify that any T-invariant of this net is a linear combination of the two minimal ones. The net in Fig. 1.6b also has two minimal T-invariants that are

$$[1, \; 0, \; 1, \; 0] \text{ and } [1, \; 1, \; 0, \; 1].$$

But, the net in Fig. 2.1a has no T-invariant.

According to the characteristic of state equation, we know that if the firing count vector of a firable transition sequence is a T-invariant, then firing this transition sequence at a marking will lead to this marking again.

Therefore, T-invariant is closely related to the liveness and boundedness of a Petri net, as shown in the following conclusion:

Theorem 2.2 ([2]) *If a Petri net* (P, T, F, M_0) *is live and bounded, then it has a T-invariant X such that*

$$||X|| = T.$$

Proof The above conclusion holds due to the fact that in a bounded and live Petri net there is a reachable marking and a firable transition sequence which contains all transitions such that firing this firable transition sequence at this marking will return this marking again, and thus this firable transition sequence corresponds to a T-invariant in which every element is a positive integer. □

Existing a T-invariant whose support is exactly the whole transition set is a necessary condition on making a Petri net be live and bounded.

2.2.2 P-Invariant

Definition 2.3 (*P-invariant*) A $|P|$-dimensional nonnegative integer vector X is a *P-invariant* of a net $N = (P, T, F)$ if the following holds:

$$X \neq \overline{0} \wedge X \cdot [N]^{\mathrm{T}} = \overline{0}$$

where $[N]^{\mathrm{T}}$ is the transpose of $[N]$.

Obviously, P-invariant and T-invariant are two dual concepts. In other words, given a net and its *dual net*, a T-invariant of this net is a P-invariant of this dual net, and vice versa.

Given a net N, if its places become transitions, its transitions become places and the directions of its arcs are unchanged, then the resulting net is its *dual net* and denoted as N^{∂}.

For example, Fig. 2.1c shows the dual net of the net in Fig. 2.1a.

Obviously, if N and N^{∂} are dual, then their incidence matrices have the following relation:

$$[N] = -[N^{\partial}]^{\mathrm{T}},$$

as shown in Fig. 2.1b, d.

Similarly, *minimal P-invariant* and *basis* can be defined, and any P-invariant can be represented by a linear combination of all minimal ones with nonnegative rational coefficients.

The *support* of a P-invariant X is defined as

$$||X|| \triangleq \{p \in P \mid X(p) > 0\}.$$

Example 2.8 The net in Fig. 2.1a has exactly two minimal P-invariants

$$[1, 1, 0, 0] \text{ and } [0, 0, 1, 1]$$

corresponding exactly to the minimal T-invariants of the net in Fig. 2.1c, but the net in Fig. 2.1c has no P-invariant. The supports of the two minimal P-invariants, respectively, are

$$\{p_1, p_2\} \text{ and } \{p_3, p_4\}.$$

P-invariant is closely related to boundedness.

For a P-invariant X of a Petri net (N, M_0), the weighted sum of tokens in the places corresponding to those non-zero elements of X are invariant for all reachable markings due to

$$M \cdot X^{\mathsf{T}} = M_0 \cdot X^{\mathsf{T}} + \psi(\sigma) \cdot [N] \cdot X^{\mathsf{T}} = M_0 \cdot X^{\mathsf{T}}$$

where σ is an arbitrary firable transition sequence and X is viewed as a weight vector corresponding to all places.

Therefore, if a Petri net has a P-invariant such that its support is exactly the whole place set, then the Petri net is bounded. In other words, existing a P-invariant in which every element is a positive integer is a sufficient condition on making a Petri net be bounded.

2.2.3 Algorithm of Computing Invariants

Since T-invariant and P-invariant are two dual concepts, we only discuss the computation of T-invariants.

Based on the definition of T-invariant, we know that the generation of T-invariants of a net is to calculate all non-trivial nonnegative integer solutions to linear equations

$$X \cdot [N] = \overline{0},$$

where a vector is *trivial* if every element of it is zero. In other words, it is to calculate a set of basic solutions of this linear equations.

Martínez and Silva proposed a simple and fast method to obtain a basis of all invariants [3]. This method is through the elementary row operations as described in Algorithm 2.1.

In this algorithm, I is a $|T| \times |T|$ identity matrix which is written on the right of $[N]$ and thus leads to a $|T| \times (|P| + |T|)$ new matrix denoted as

---

Algorithm 2.1: Computing a Basis of T-invariants of a Net.

Input: The incidence matrix of a net N;
Output: A matrix B whose row vectors form a basis of T-invariants of N;
1   $B \leftarrow [N \mid I]$;
2   **for** $j \leftarrow 1$ *to* $|P|$ **do**
3   Insert into B every new row constructed by a positive coefficient linear combination of any two rows of the original matrix B such that the j-th element of the new row is zero;
4   Delete from B any row whose j-th element is nonzero;
5   Delete the first $|P|$ columns from B;

$$[N \mid I].$$

Through the elementary row operations to this matrix named as B in this algorithm, a basis of T-invariants of N can be produced.

Specifically speaking, for the jth column of the current matrix B where j is from 1 to $|P|$, if two rows, say k_1 and k_2, satisfy

$$B(k_1, j) \cdot B(k_2, j) < 0,$$

then we can obtain a new vector through the calculation

$$\frac{|B(k_2, j)|}{\gcd\left(|B(k_1, j)|, |B(k_2, j)|\right)} \cdot B(k_1) + \frac{|B(k_1, j)|}{\gcd\left(|B(k_1, j)|, |B(k_2, j)|\right)} \cdot B(k_2)$$

where the function gcd means the greatest common divisor of two integers.

Obviously, the jth element of the new vector is zero and the vector is appended to B as a new row. After such row pairs are conducted (for the jth column), all old rows whose jth elements are nonzero are deleted from B. For the resulting B, we continue to conduct its next column (i.e., $j++$) until the $|P|$th column is done. Finally, we delete the first $|P|$ columns of B and thus all row vectors form a basis of T-invariants of N.

Through the net in Fig. 2.1c, we illustrate this computation procedure.

Example 2.9 The incidence matrix of the net in Fig. 2.1c is shown in Fig. 2.1d and thus we have $[N \mid I]$ as follows:

$$\begin{bmatrix} 1 & 0 & 1 & 0 & 0 & 0 \\ -1 & 0 & 0 & 1 & 0 & 0 \\ 1 & -1 & 0 & 0 & 1 & 0 \\ -1 & 1 & 0 & 0 & 0 & 1 \end{bmatrix}.$$

For the first column of the above matrix, since the signs of the first elements of the first and second rows are different, we obtain a new vector by the following calculation

$$1 \cdot [1, 0, 1, 0, 0, 0] + 1 \cdot [-1, 0, 0, 1, 0, 0] = [0, 0, 1, 1, 0, 0].$$

Similarly, for the first and fourth rows, the second and third rows, and the third and fourth rows, we also obtain the following three new vectors

$$[0, 1, 1, 0, 0, 1],$$

$$[0, -1, 0, 1, 1, 0],$$

$$[0, 0, 0, 0, 1, 1].$$

The above four new vectors are inserted into the above matrix as new rows, and then the four old rows are deleted since their first elements are all nonzero. Then, we obtain the following matrix

$$\begin{bmatrix} 0 & 0 & 1 & 1 & 0 & 0 \\ 0 & 1 & 1 & 0 & 0 & 1 \\ 0 & -1 & 0 & 1 & 1 & 0 \\ 0 & 0 & 0 & 0 & 1 & 1 \end{bmatrix}.$$

For the second column of the above matrix, we only need to perform a linear combination for the second and third rows since the signs of the related two elements are different; and we thus obtain a new row vector:

$$[0, 0, 1, 1, 1, 1].$$

After inserting this new row and deleting the two old ones, we will then obtain the following matrix:

$$\begin{bmatrix} 0 & 0 & 1 & 1 & 0 & 0 \\ 0 & 0 & 0 & 0 & 1 & 1 \\ 0 & 0 & 1 & 1 & 1 & 1 \end{bmatrix}.$$

Finally, we obtain three vectors

$$[1, 1, 0, 0], \quad [0, 0, 1, 1] \text{ and } [1, 1, 1, 1]$$

that form a basis of T-invariants of the net. Certainly, $[1, 1, 1, 1]$ can be removed out of this basis since it can be represented by another two.

2.3   Repetitiveness

This section introduces the concept of repetitive vector as well as a relation between repetitive vectors and liveness. We also illustrate a relation between repetitive vector and T-invariant that can be used to compute a basis of all repetitive vectors of a net. More details can be found in [1, 4].

2.3.1   Repetitive Vector

Definition 2.4 (*Repetitive Vector*) A $|T|$-dimensional nonnegative integer vector X is a *repetitive vector* of a net $N = (P, T, F)$ if the following holds:

$$X \neq \overline{0} \wedge X \cdot [N] \geq \overline{0}.$$

According to Definitions 2.2 and 2.4, a T-invariant is a special repetitive vector.

Given a repetitive vector, if any less one is not a repetitive vector, then it is called a *minimal repetitive vector*.

Given a net and k repetitive vectors $X_1, \cdots$, and X_k of it, if any repetitive vector of it can be represented by a linear combination of $X_1, \cdots$, and X_k with nonnegative rational coefficients, then $\{X_1, \ldots, X_k\}$ is called a *basis* of all repetitive vectors of this net. The *support* of a repetitive vector X of a net is defined as

$$||X|| \triangleq \{t \in T \mid X(t) > 0\}.$$

A repetitive vector is not necessarily represented by a linear combination of all minimal ones.

Example 2.10   The net in Fig. 2.1c has exactly two minimal repetitive vector

$$[1, 0, 0, 0] \text{ and } [0, 0, 1, 1],$$

but the repetitive vector $[1, 1, 1, 1]$ cannot be represented by them. In fact, the following three ones form a basis:

$$[1, 0, 0, 0], \quad [1, 1, 0, 0] \text{ and } [0, 0, 1, 1].$$

Repetitiveness is closely related to liveness.

According to the state equation of a net, we know that if a firable transition sequence corresponds to a repetitive vector, then the marking reached by firing this transition sequence covers the original one and thus this transition sequence can be fired repeatedly.

Definition 2.5 (*Structural Repetitiveness*) A (Petri) net is *structurally repetitive* if there exists a repetitive vector X such that

$$||X|| = T$$

where T is the transition set of this (Petri) net.

Theorem 2.3 ([5–7]) *If a Petri net is live, then it is structurally repetitive.*

Proof When a Petri net (N, M_0) is live, there must be a firable transition sequence σ_0 such that

$$\psi(\sigma_0) > \bar{0},$$

i.e., each element of σ_0 is a positive integer. We let $M_0[\sigma_0\rangle M_1$ and thus (N, M_1) is still live. Hence, (N, M_1) also has a firable transition sequence σ_1 such that

$$\psi(\sigma_1) > \bar{0}.$$

Similarly, (N, M_2) is still live where $M_1[\sigma_1\rangle M_2$. We repeat the above process and thus have an infinite sequence

$$M_0[\sigma_0\rangle M_1[\sigma_1\rangle M_2[\sigma_2\rangle \cdots$$

such that

$$\forall j \in \mathbb{N} : \psi(\sigma_j) > \bar{0}.$$

According to Dickson's Lemma [8], there exist two markings M_j and M_k in the above infinite sequence such that

$$j < k \wedge M_j < M_k \wedge M_j[\sigma_j\sigma_{j+1}\cdots\sigma_{k-1}\rangle M_k.$$

Obviously,

$$\psi(\sigma_j \sigma_{j+1} \cdots \sigma_{k-1}) > \overline{0} \wedge \psi(\sigma_j \sigma_{j+1} \cdots \sigma_{k-1}) \cdot [N] = M_k - M_j \geq \overline{0},$$

i.e., $\psi(\sigma_j \sigma_{j+1} \cdots \sigma_{k-1})$ is a repetitive vector in which each element is a positive integer. $\square$

Question

When referring to the definition of repetitive vector, we may require that a $|P|$-dimensional nonnegative integer vector X satisfies

$$X \neq \overline{0} \wedge [N] \cdot X \geq \overline{0}$$

or

$$X \neq \overline{0} \wedge [N] \cdot X \leq \overline{0}.$$

Obviously, P-invariant is a special case of this definition, but what are properties brought by such a vector? Try to analyse it.

2.3.2   Relation Between Repetitive Vector and T-Invariant

2.3.2.1   Transition-Added Net

Although we know such a simple relation that a T-invariant is a repetitive vector in a net, a more useful relation between them is shown here. The idea behind this relation comes from the following fact.

> If a firable transition sequence corresponds to a repetitive vector, then after firing the transition sequence at a given marking will yields a new marking that is equal to or greater than the given one. Therefore, if we remove those new added tokens out of the related places, then the resulting marking becomes the given one, which is obviously a behaviour caused by a T-invariant.

Based on this idea, we construct the following new net for a given net:

Definition 2.6 (*Transition-added Net*) Given a net $N = (P, T, F)$, the net $N^a \triangleq (P, T \cup T', F \cup F')$ is the *transition-added net* of N if the following conditions hold:

1. $T \cap T' = \emptyset \wedge |T'| = |P| \wedge \forall t \in T' : |{}^\bullet t| = 1$,

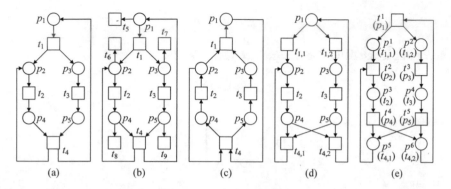

Fig. 2.2 **a** A simple net coming from [7]; **b** the transition-added net of the net in **a**; **c** the inverse net of the net in **a**; **d** the transition-split net of the net in **a**; and **e** the dual net of the net in **d**

2. $\forall p \in P, \exists t \in T': (p, t) \in F'$, and
3. $(p, t) \in F' \Rightarrow p \in P \wedge t \in T'$.

In fact, the transition-added net N^a is produced via augmenting a new output transition to each place of N. The relation of their incidence matrices is

$$[N^a] = \begin{bmatrix} N \\ -I \end{bmatrix}.$$

Example 2.11 The net in Fig. 2.2b shows the transition-added net of the net in (a). At such a marking $\{p_1\}$ of (a), firing the transition sequence $t_1 t_2 t_3 t_4$ will yield the marking $\{p_1, p_2\}$. Hence, if we remove this token out of p_2, then the original marking is restored.

2.3.2.2 A Relation Between a Net and Its Transition-Added Net

Next we show that there exists a one-to-one correspondence between the repetitive vectors of a net and the T-invariants of its transition-added net.

Theorem 2.4 ([4]) *A nonnegative integer vector X is a repetitive vector of a net N if and only if* $[X, Y]$ *is a T-invariant of* N^a *where*

$$X \neq \overline{0} \ and \ Y = X \cdot [N].$$

Proof The sufficiency obviously holds since the following holds:

$$X \neq \overline{0} \wedge X \cdot [N] = Y \geq \overline{0}.$$

(Necessity) If X is a repetitive vector of a net N, we have

$$X \cdot [N] \geq \overline{0}$$

according to the definition of repetitive vector. We let

$$Y = X \cdot [N].$$

In consequence, we have

$$[X, Y] \cdot \begin{bmatrix} N \\ -I \end{bmatrix} = \overline{0}.$$

Because

$$\begin{bmatrix} N \\ -I \end{bmatrix}$$

is the incidence matrix of N^a, $[X, Y]$ is a T-invariant of N^a.   □

Based on Theorem 2.4, we can draw the following conclusion:

Theorem 2.5 ([4]) *If $\{X_1, \ldots, X_k\}$ is a basis of T-invariants of N^a, then $\{X_1 \upharpoonright T, \ldots, X_k \upharpoonright T\}$ is a basis of repetitive vectors of N, where T is the transition set of N, the first $|T|$ elements of each X_j are indexed by transitions of T, and $X_j \upharpoonright T$ means the projection of X_j on T, $\forall j \in \mathbb{N}_k^+$.*

Notice that given a vector X indexed by T and given a subset T' of T, the *projection* of X on T' is defined as a new vector X' indexed by T' such that

$$\forall t \in T' : X'(t) = X(t).$$

For convenience, this projection X' is also denoted by $X \upharpoonright T'$.

2.3.3 Algorithm of Computing Repetitive Vectors

2.3.3.1 A General Procedure

Based on Theorems 2.4 and 2.5, the computation of a basis of repetitive vectors of a net can be transferred into the computation of a basis of T-invariants of its transition-added net.

We may first construct the transition-added net of a net, then compute the T-invariants of this transition-added net, and finally compute the projections of these T-invariants on the transition set of the original net. Certainly, all these can be operated over their matrices.

Example 2.12 The examples in Fig. 2.2a, b are used to show the computing process. The incidence matrices of the net in Fig. 2.2a and its transition-added net in Fig. 2.2b, respectively, are

$$\begin{bmatrix} -1 & 1 & 1 & 0 & 0 \\ 0 & -1 & 0 & 1 & 0 \\ 0 & 0 & -1 & 0 & 1 \\ 1 & 1 & 0 & -1 & -1 \end{bmatrix}$$

and

$$\begin{bmatrix} -1 & 1 & 1 & 0 & 0 \\ 0 & -1 & 0 & 1 & 0 \\ 0 & 0 & -1 & 0 & 1 \\ 1 & 1 & 0 & -1 & -1 \\ -1 & 0 & 0 & 0 & 0 \\ 0 & -1 & 0 & 0 & 0 \\ 0 & 0 & -1 & 0 & 0 \\ 0 & 0 & 0 & -1 & 0 \\ 0 & 0 & 0 & 0 & -1 \end{bmatrix}.$$

After taking the second matrix as the input of Algorithm 2.1, we can obtain a basis of T-invariants of the net in Fig. 2.2b:

$$[1, 2, 1, 1, 0, 0, 0, 1, 0] \text{ and } [2, 2, 2, 2, 0, 2, 0, 0, 0].$$

According to Theorem 2.5, we hence obtain a basis of repetitive vectors of the net in Fig. 2.2a:

$$[1, 2, 1, 1] \text{ and } [2, 2, 2, 2].$$

2.3.3.2 An Improved Procedure

Can we compute a basis of repetitive vectors of a given net directly according to its incidence matrix?

From the above process of computing a basis of repetitive vectors, we easily see that when the matrix

$$\begin{bmatrix} N \\ -I \end{bmatrix}$$

is taken as the input of Algorithm 2.1, the function of every -1 of $[-I]$ is only to change those positive integers that are in the same column with this -1 into zero and to make these resulting rows be saved.

Based on the above fact, we can obtain an algorithm (Algorithm 2.2) of computing a basis of repetitive vectors of a net through modifying Line 4 of Algorithm 2.1:

Algorithm 2.2: Computing a Basis of Repetitive Vectors of a Net.

Input: The incidence matrix of a net N;
Output: A matrix B whose row vectors form a basis of repetitive vectors of N;
1 $B \leftarrow [N \mid I]$;
2 **for** $j \leftarrow 1$ *to* $|P|$ **do**
3 $\quad$ Insert into B every new row constructed by a positive coefficient linear combination of
   $\quad$ any two rows of the original matrix B such that the j-th element of the new row is zero;
4 $\quad$ Delete from B any row whose j-th element is negative;
5 Delete the first $|P|$ columns from B;

Delete from B any row whose jth element is negative replaces *Delete from B any row whose jth element is nonzero*. In other words, we save those rows in which the jth element is nonnegative.

This algorithm directly takes the incidence matrix of a net as its input and outputs a basis of repetitive vectors of this net.

Example 2.13 The net in Fig. 2.2a is used to show this computing process. The incidence matrix of this net is directly input into Algorithm 2.2 and then the matrix $[N \mid I]$ is obtained as follows:

$$\begin{bmatrix} -1 & 1 & 1 & 0 & 0 & 1 & 0 & 0 & 0 \\ 0 & -1 & 0 & 1 & 0 & 0 & 1 & 0 & 0 \\ 0 & 0 & -1 & 0 & 1 & 0 & 0 & 1 & 0 \\ 1 & 1 & 0 & -1 & -1 & 0 & 0 & 0 & 1 \end{bmatrix}.$$

For the first column of the above matrix, we only need to combine the first and fourth rows and thus obtain a new row as follows:

$$[0 \quad 2 \quad 1 \quad -1 \quad -1 \quad 1 \quad 0 \quad 0 \quad 1].$$

As a result, this new row is inserted into this matrix, and only the first row should be deleted since the first element of this row is less than 0. The following matrix is thus obtained:

$$\begin{bmatrix} 0 & -1 & 0 & 1 & 0 & 0 & 1 & 0 & 0 \\ 0 & 0 & -1 & 0 & 1 & 0 & 0 & 1 & 0 \\ 1 & 1 & 0 & -1 & -1 & 0 & 0 & 0 & 1 \\ 0 & 2 & 1 & -1 & -1 & 1 & 0 & 0 & 1 \end{bmatrix}.$$

For the second column of the above matrix, we should combine the first and third rows as well as the first and fourth rows, and then delete the first row. The following matrix is produced:

$$\begin{bmatrix} 0 & 0 & -1 & 0 & 1 & 0 & 0 & 1 & 0 \\ 1 & 1 & 0 & -1 & -1 & 0 & 0 & 0 & 1 \\ 0 & 2 & 1 & -1 & -1 & 1 & 0 & 0 & 1 \\ 1 & 0 & 0 & 0 & -1 & 0 & 1 & 0 & 1 \\ 0 & 0 & 1 & 1 & -1 & 0 & 2 & 0 & 1 \end{bmatrix}.$$

Similarly, we repeat the above process and obtain the final matrix as follows:

$$\begin{bmatrix} 0 & 0 & 0 & 1 & 0 & 1 & 2 & 1 & 1 \\ 0 & 2 & 0 & 0 & 0 & 2 & 2 & 2 & 2 \end{bmatrix}.$$

After deleting the five columns, we obtain the following two vectors:

$$[1, 2, 1, 1] \text{ and } [2, 2, 2, 2]$$

that exactly form a basis of repetitive vectors of the net in Fig. 2.2a.

2.4 Siphon and Trap

This section first introduces the concepts of siphon and trap that are defined from the perspective of places and based on the inclusion relation of sets. As we know, T-invariant and repetitive vector are defined from the perspective of transitions and based on the algebra calculation; however, a relation between siphons and repetitive vectors is discovered and can be utilised to compute siphons and traps. And this computation procedure of siphons and traps is based on elementary row operations of incidence matrices. More details can be found in literatures [1, 4].

2.4.1 Siphon

Definition 2.7 (*Siphon*) A nonempty set S is a *siphon* of a net $N = (P, T, F)$ if the following holds:

$$S \subseteq P \wedge {}^{\bullet}S \subseteq S^{\bullet}.$$

If any proper subset of a siphon is not a siphon, then the siphon is *minimal*.

Given a collection $\{S_1, \ldots, S_k\}$ of siphons of a net, if any siphon of the net can be represented by the union of some siphons in $\{S_1, \ldots, S_k\}$, then $\{S_1, \ldots, S_k\}$ is a *basis* of all siphons of the net.

The union of any two siphons is still a siphon in a net. But, all minimal siphons do not necessarily form a basis.

Example 2.14 The net in Fig. 2.2a has 5 siphons:

$$\{p_1, p_2, p_4\}, \{p_1, p_3, p_5\},$$
$$\{p_1, p_2, p_3, p_4\}, \{p_1, p_2, p_3, p_5\}, \{p_1, p_2, p_3, p_4, p_5\}$$

where the minimal ones are $\{p_1, p_2, p_4\}$ and $\{p_1, p_3, p_5\}$ that are not a basis. The first four siphons form a basis.

A siphon will never be marked once tokens are all removed out of it, due to

$$^{\bullet}S \subseteq S^{\bullet}.$$

Therefore, siphons are closely related to the liveness of a Petri net since all transitions associated with an unmarked siphon are never enabled.

Theorem 2.6 ([9]) *If a Petri net is live, then every siphon is marked at every reachable marking.*

Theorem 2.7 ([9]) *If a Petri net has a deadlock, then it has a siphon that is emptied at the deadlock.*

Proof When a deadlock occurs, every transition becomes disabled. Hence, we can select an arbitrary transition, and then it must has an input place that is unmarked at the deadlock. Since every input transition of this place is disabled either at the deadlock, each of these input transitions also has an input place that is unmarked at the deadlock. Due to the finiteness of places and transitions, we can repeat the above selection process and finally find a set of places satisfying that *they are unmarked at the deadlock and their input transitions are also their outputs*, i.e., they form a siphon that has no token at this deadlock.   □

In other words, all places unmarked by a deadlock form a siphon.

2.4.2 Trap

Definition 2.8 (*Trap*) A nonempty set Q is a *trap* of a net $N = (P, T, F)$ if the following holds:

$$Q \subseteq P \wedge Q^{\bullet} \subseteq {}^{\bullet}Q.$$

Trap and siphon are a pair of dual concepts. A trap of a net is exactly a siphon of the *inverse net* of this net, and vice versa.

The *inverse net* of a net N is produced by changing the directions of all arcs of N [10] and denoted as N^{iv}.

Obviously, there is the following relation between the incidence matrices of a net N and its inverse net N^{iv}:

$$[N] = -[N^{iv}].$$

For example, Fig. 2.2c shows the inverse net of (a).

If any proper subset of a trap is not a trap, this trap is *minimal*.

Given a collection $\{Q_1, \ldots, Q_k\}$ of traps of a net, if any trap can be represented by the union of some traps in $\{Q_1, \ldots, Q_k\}$, then $\{Q_1, \ldots, Q_k\}$ is a *basis* of all traps of the net. The union of any two traps is still a trap in a net.

Example 2.15 The net in Fig. 2.2a has the following 8 traps:

$$\{p_2, p_4\}, \{p_1, p_2, p_4\}, \{p_2, p_4, p_5\}, \{p_1, p_3, p_5\},$$

$$\{p_1, p_3, p_4, p_5\}, \{p_2, p_3, p_4, p_5\}, \{p_1, p_2, p_4, p_5\}, \{p_1, p_2, p_3, p_4, p_5\}$$

in which $\{p_2, p_4\}$ and $\{p_1, p_3, p_5\}$ are minimal and the first six form a basis. They are exactly the siphons of the net in Fig. 2.2c. The nets in Fig. 2.2a, c are mutually inverse to each other.

A trap will be marked forever once it is marked, due to

$$Q^\bullet \subseteq {}^\bullet Q.$$

Therefore, traps are also closely related to liveness. In the next chapter, it will be seen that trap and siphon can be used to decide the liveness of some Petri nets.

2.4.3 Relation Between Siphon and Repetitive Vector

Since trap and siphon are a pair of dual concepts, we only consider siphon here.

2.4.3.1   Transition-Split Net

It seems impossible that there is a close relation between siphon and repetitive vector since the former is a concept related to places while the latter is a concept related to transitions. But such a relation can indeed be built through a so-called *transition-split net*.

Definition 2.9 (*Transition-split Net*) Given a net $N = (P, T, F)$, its *transition-split net* $N^s \triangleq (P', T', F')$ is defined as:

1. $P' = P$; and
2. T' and F' can be generated by considering each $t \in T$ (for convenience, we denote $t^{\bullet} = \{p_1, \ldots, p_k\}$) as follows:

    a. if $k \leq 1$, a new transition is put into T' such that its pre-set and post-set are the same with t's; or
    b. if $k > 1$, k new transitions, say $t_1, \ldots, t_k$, are put into T' such that

$$\forall j \in \mathbb{N}_k^+ : {}^{\bullet}t_j = {}^{\bullet}t \wedge t_j^{\bullet} = \{p_j\}.$$

Example 2.16   The net in Fig. 2.2d is the transition-split net of the net in Fig. 2.2a where the transition t_1 (respectively t_4) is split into two transitions $t_{1,1}$ and $t_{1,2}$ (respectively $t_{4,1}$ and $t_{4,2}$) since it has two output places.

2.4.3.2   A Relation Between a Net and Its Transition-Split Net

Theorem 2.8 ([4]) *A net N and its transition-split net N^s have the same siphons.*

Proof   On the one hand, we let S be a siphon of N and t be an arbitrary transition in N^s such that

$$t \in {}^{\bullet}S.$$

And we let t' be the transition in N such that t is produced through splitting t'. According to Definition 2.9 we thus have that

$$ {}^{\bullet}t = {}^{\bullet}t' \wedge t^{\bullet} \subseteq t'^{\bullet}.$$

Hence, $t' \in {}^{\bullet}S$ in N. Hence, $t' \in S^{\bullet}$ in N since S is a siphon of N. Hence, $t \in S^{\bullet}$ in N^s since ${}^{\bullet}t = {}^{\bullet}t'$, which means that S is also a siphon of N^s.

The above proof is easily understood through the siphon $\{p_1, p_2, p_4\}$ of the net in Fig. 2.2a and the transitions $t_{4,2}$ in Fig. 2.2d (corresponding to t in the above proof) and t_4 in Fig. 2.2a (corresponding to t' in the above proof).

On the other hand, we let S be a siphon of N^s and t be an arbitrary transition in N such that

$$t \in {}^\bullet S.$$

Since t is split into a group of transitions while generating N^s, N^s has at least one transition, say t', corresponding to t and satisfying $t' \in {}^\bullet S$ in N^s. Hence, $t' \in S^\bullet$ in N^s since S is a siphon of N^s. Since the pre-set of t' in N^s is equal to the pre-set of t in N, we have that $t \in S^\bullet$ in N, which means that S is also a siphon of N. □

The proof of the second part is easily understood through the siphon $\{p_1, p_2, p_4\}$ of the net in Fig. 2.2d and the transitions t_1 in Fig. 2.2a (corresponding to t in the above proof) and $t_{1,1}$ in Fig. 2.2d (corresponding to t' in the above proof).

Example 2.17 We can verify that the nets in Fig. 2.2a, d have the same siphons as follows:

$$\{p_1, p_2, p_4\}, \quad \{p_1, p_3, p_5\},$$
$$\{p_1, p_2, p_3, p_4\}, \quad \{p_1, p_2, p_3, p_5\}, \quad \{p_1, p_2, p_3, p_4, p_5\}.$$

2.4.3.3 A Relation Between a Transition-Split Net and Its Dual Net

Next, we show that for such a net in which each transition has one output place at most, there is a one-to-one correspondence relation between its siphons and the supports of repetitive vectors of its dual net.

Theorem 2.9 ([4]) *Given a net N satisfying $|t^\bullet| \le 1$ for each transition t and given its dual net N^∂, then for each siphon of N, there is a repetitive vector of N^∂ such that its support corresponds to this siphon; and vice versa.*

Proof We first observe that each place of N^∂ has at most one output transition since each transition of N has at most one output place. Therefore, there is at most one -1 in each column of the incidence matrix $[N^\partial]$. We let S be a siphon of N. The set of those transitions of N^∂ corresponding to S are denoted as T_S. Due to

$$ {}^\bullet S \subseteq S^\bullet \text{ in } N, $$

we have

$$^\bullet T_S \subseteq T_S^\bullet \text{ in } N^\partial.$$

We construct a vector X indexed by all transitions of N^∂ as follows:

$$X(t) = \begin{cases} 1 & t \in T_S \\ 0 & t \notin T_S \end{cases}.$$

We obviously have

$$S = ||X||.$$

Now we illustrate that X is a repetitive vector of N^∂ by proving

$$X \cdot [N^\partial](p) \geq 0$$

for each place p of N^∂, where $[N^\partial](p)$ is the column vector of incidence matrix $[N^\partial]$ corresponding to p. We consider the following two cases:

– In case of $p \in {}^\bullet T_S$, then there exists one and only one -1 in $[N^\partial](p)$ and this -1 corresponds to a transition in T_S (i.e., this transition is the output of p). Additionally, we have

$$p \in T_S^\bullet$$

due to

$$^\bullet T_S \subseteq T_S^\bullet.$$

Therefore, there is at least one 1 in $[N^\partial](p)$ such that this 1 corresponds to a transition in T_S (i.e., this transition is the input of p). Therefore, we have

$$X \cdot [N^\partial](p) \geq 0.$$

– In case of $p \notin {}^\bullet T_S$, then in $[N^\partial](p)$, those elements corresponding to the transitions in T_S are all nonnegative. Therefore, we also have

$$X \cdot [N^\partial](p) \geq 0.$$

In a word, for each siphon of N, there exists a repetitive vector of N^∂ whose support corresponds to this siphon.

(Vice Versa) We let X be a repetitive vector of N^∂ and denote $S_{||X||}$ as the set of places corresponding to $||X||$ in N. We let t be an arbitrary transition of N such that

$$t \in {}^\bullet S_{||X||},$$

and denote p_t as the place of N^∂ corresponding to t. Then, we obviously have

$$p_t \in {}^\bullet ||X||.$$

Hence, there is a transition t' of N^∂ such that

$$X(t') > 0 \wedge [N^\partial](t', p_t) = -1.$$

Additionally, because of

$$X \cdot [N^\partial](p_t) \geq 0,$$

there is a set of transitions of N^∂, say $\{t_1, \ldots, t_k\}$, satisfying the following conditions:

$$X(t') \cdot [N^\partial](t', p_t) + \sum_{j=1}^{k} X(t_j) \cdot [N^\partial](t_j, p_t) \geq 0$$

and

$$\forall j \in \mathbb{N}_k^+ : X(t_j) > 0 \wedge [N^\partial](t_j, p_t) = 1.$$

In other words, the place p_t is an output of t_j for each $j \in \mathbb{N}_k^+$. Hence, we have

$$p_t \in ||X||^\bullet \text{ in } N^\partial.$$

Hence, we have

$$t \in S^\bullet_{||X||} \text{ in } N.$$

Hence, $S_{||X||}$ is a siphon of N. In a word, for each repetitive vector of N^∂, its support corresponds to a siphon of the original net N. $\qquad\square$

Example 2.18 $\{p_1, p_2, p_4\}$ is a siphon of the net in Fig. 2.2d, and the constructed vector indexed by the transitions of the net in Fig. 2.2e is

$$\begin{array}{ccccc} t^1 & t^2 & t^3 & t^4 & t^5 \\ [1, & 1, & 0, & 1, & 0]. \end{array}$$

It is multiplied by the incidence matrix of the net in Fig. 2.2e and then we have the following result:

$$\begin{array}{c} t^1\ t^2\ t^3\ t^4\ t^5 \\ [1, 1, 0, 1, 0] \end{array} \cdot \begin{array}{c} p^1\ p^2\ p^3\ p^4\ p^5\ p^6 \\ \begin{bmatrix} 1 & 1 & 0 & 0 & 0 & -1 \\ -1 & 0 & 1 & 0 & -1 & 0 \\ 0 & -1 & 0 & 1 & 0 & 0 \\ 0 & 0 & -1 & 0 & 1 & 1 \\ 0 & 0 & 0 & -1 & 1 & 1 \end{bmatrix} \begin{array}{c} t^1 \\ t^2 \\ t^3 \\ t^4 \\ t^5 \end{array} \end{array} = [0, 1, 0, 0, 0].$$

Hence, the constructed vector is a repetitive vector of the net in Fig. 2.2e. On the other hand,

$$\begin{array}{ccccc} t^1 & t^2 & t^3 & t^4 & t^5 \\ X = [1, & 1, & 0, & 1, & 0] \end{array}$$

Algorithm 2.3: Computing a Basis of Siphons of a Net.

Input: A net N;
Output: A basis of siphons of N;
1 Generate the transition-split net N^s of N;
2 Generate the dual net (denoted as $N^{s \cdot \partial}$) of N^s;
3 Compute a basis of repetitive vectors of $N^{s \cdot \partial}$ by Algorithm 2.2;
4 Compute the supports of these repetitive vectors;
5 The place sets corresponding to these supports are a basis of siphons of N;

---

is a repetitive vector of the net in Fig. 2.2e, $\{p_1, p_2, p_4\}$ is the place set of the net in Fig. 2.2d corresponding to the support of X, and $t_{4,2}$ is an input transition of this place set. We can see that in the net in Fig. 2.2e, the place p^6 corresponds to $t_{4,2}$. Therefore, there is a transition in the net in Fig. 2.2e, i.e. t^1, such that

$$X(t^1) > 0 \land [N^\partial](t^1, p^6) = -1.$$

Consequently, we can find the transition t^4 that guarantees

$$X \cdot [N^\partial](p^6) > 0,$$

i.e., t^4 in $||X||$ is an input of p^6. In consequence, we know that in the net in Fig. 2.2d, the place p_4 corresponding to t^4 is an input of $t_{4,2}$. In other words, $t_{4,2}$ is an output of $\{p_1, p_2, p_4\}$.

2.4.4 Algorithm of Computing Siphon and Trap

Based on the relation discovered in the above section, we can design an algorithm to generate a basis of siphons of a net N, as shown in Algorithm 2.3.

In fact, each step of this computing procedure can be performed over their incidence matrices. The incidence matrix of the dual net can be generated according to the following formula:

$$[N^{s \cdot \partial}] = -[N^s]^\mathsf{T},$$

while the incidence matrix of the transition-split net N^s can be generated through the incidence matrix $[N]$ as shown in Algorithm 2.4. In this algorithm, if a row of $[N]$ has multiple 1's (i.e., the related transition has multiple output places), then multiple copies of the row are produced and every copy maintains only one 1 while other 1's are replaced with 0's.

Example 2.19 In the incidence matrix of the net in Fig. 2.2a, the row vector corresponding to the transition t_1 has two 1's. Hence, two new row vectors are produced

---
Algorithm 2.4: Computing the Incidence Matrix of the Transition-Split Net of a Net.

Input: The incidence matrix of a net N;
Output: The incidence matrix of the transition-siplit net N^s;
1 $[N^s] \leftarrow \emptyset$;
2 **for** *each transition t of N* **do**
3   **if** *[N](t) has no or only one 1* **then**
4          $X \leftarrow [N](t)$;
5   Insert X into $[N^s]$;
6   **else**
7   **for** *each p with [N](t, p) = 1* **do**
8              $X \leftarrow [N](t)$;
9   Replace other 1's with 0's except the 1 corresponding to p in X;
10   Insert X into $[N^s]$;

---

through the following splitting:

$$
t_1 \begin{array}{c} \begin{array}{ccccc} p_1 & p_2 & p_3 & p_4 & p_5 \end{array} \\ \left[\begin{array}{ccccc} -1 & 1 & 1 & 0 & 0 \end{array} \right] \end{array}
\longmapsto
\begin{array}{c} \begin{array}{ccccc} p_1 & p_2 & p_3 & p_4 & p_5 \end{array} \\ \begin{array}{c} t_{1,1} \\ t_{1,2} \end{array} \left[\begin{array}{ccccc} -1 & 1 & 0 & 0 & 0 \\ -1 & 0 & 1 & 0 & 0 \end{array} \right] \end{array}.
$$

Other splittings can be conducted accordingly.

Example 2.20 According to Algorithm 2.2, we can compute a basis of repetitive vectors of the net in Fig. 2.2e:

$$[1, 1, 0, 1, 0]$$

$$[1, 0, 1, 0, 1]$$

$$[1, 1, 1, 0, 1]$$

$$[1, 1, 1, 1, 0]$$

$$[1, 1, 1, 1, 1]$$

$$[2, 1, 1, 1, 1]$$

$$[2, 2, 1, 1, 1]$$

in which the entries are indexed by t^1, t^2, t^3, t^4 and t^5. The supports of the seven repetitive vectors are

$$\{t^1, \ t^2, \ t^4\}, \ \{t^1, \ t^3, \ t^5\},$$

$$\{t^1, \quad t^2, \quad t^3, \quad t^4\}, \quad \{t^1, \quad t^2, \quad t^3, \quad t^5\}, \quad \{t^1, \quad t^2, \quad t^3, \quad t^4, \quad t^5\}$$

which correspond to the following place sets of the net in Fig. 2.2a:

$$\{p_1, \quad p_2, \quad p_4\}, \quad \{p_1, \quad p_3, \quad p_5\},$$

$$\{p_1, \quad p_2, \quad p_3, \quad p_4\}, \quad \{p_1, \quad p_2, \quad p_3, \quad p_5\}, \quad \{p_1, \quad p_2, \quad p_3, \quad p_4, \quad p_5\}$$

that are exactly all the siphons of this net.

2.5   Application

We continue to discuss and analyse the dining philosophers problem on the basis of the structure concepts introduced in this chapter. Fig. 1.1 in Chap. 1 shows a Petri net modelling this problem, and let's first observe the reason causing a deadlock in this model.

Reason of Deadlock

From the aspect of its physical background, the reason behind its deadlock is that each philosopher holds her/his left chopstick but is waiting for the right one, so that everyone cannot hold the related two chopsticks to eat.

From the aspect of its Petri net model, the reason behind the deadlock is that the model has such a (minimal) siphon

$$\{p_{j,3}, \ r_j \mid j \in \mathbb{N}_5^+\},$$

as shown in Fig. 2.3, which is emptied so that all transitions become disabled. From Fig. 2.3 it is easy to see that just transitions $t_{j,1}$ ($j \in \mathbb{N}_5^+$) remove the tokens from this siphon.

Figure 1.12 in Chap. 1 shows a deadlock-free model: the two chopsticks needed by a philosopher are allocated to her/him at one time when s/he is ready for eating. We can obtain all minimal siphons of it:

$$\{p_{j,3}, \ r_j \mid j \in \mathbb{N}_5^+\} \text{ and } \{p_{j,1}, \ p_{j,2}, \ p_{j,3}\}, \quad \forall j \in \mathbb{N}_5^+.$$

Fig. 2.3 The siphon resulting in the deadlock of the Petri net in Fig. 1.1. Here shows the subnet associated with the siphon through which the reason resulting in the deadlock can be seen clearly

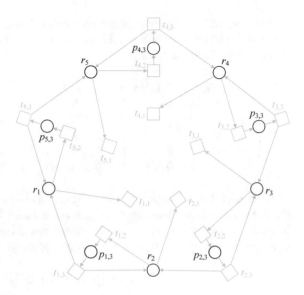

And we can verify that each of them is a trap. This means that each siphon can never be emptied so that no deadlock occurs in it.

Starvation

However, this deadlock-free solution results in another problem: *starvation*.

If a philosopher is ready for eating but one of her/his neighbours has been

thinking → ready for eating → holding two chopsticks and eating → thinking → · · ·

again and again, then this philosopher is in a starvation situation. This is an *unfair* phenomenon:

A process has always been suspended, but it has not been awaken since the resources it needs has not been allocated to it.

These starvation phenomena can also be analysed from the aspect of this Petri net model. The Petri net in Fig. 1.12 has five basic T-invariants as shown by the row vectors of the following matrix:

$$
\begin{array}{ccccccccccccccc}
t_{1,1} & t_{1,2} & t_{1,3} & t_{2,1} & t_{2,2} & t_{2,3} & t_{3,1} & t_{3,2} & t_{3,3} & t_{4,1} & t_{4,2} & t_{4,3} & t_{5,1} & t_{5,2} & t_{5,3} \\
\left[\begin{array}{ccccccccccccccc}
1 & 1 & 1 & 0 & 0 & 0 & 0 & 0 & 0 & 0 & 0 & 0 & 0 & 0 & 0 \\
0 & 0 & 0 & 1 & 1 & 1 & 0 & 0 & 0 & 0 & 0 & 0 & 0 & 0 & 0 \\
0 & 0 & 0 & 0 & 0 & 0 & 1 & 1 & 1 & 0 & 0 & 0 & 0 & 0 & 0 \\
0 & 0 & 0 & 0 & 0 & 0 & 0 & 0 & 0 & 1 & 1 & 1 & 0 & 0 & 0 \\
0 & 0 & 0 & 0 & 0 & 0 & 0 & 0 & 0 & 0 & 0 & 0 & 1 & 1 & 1
\end{array}\right].
\end{array}
$$

Here, each basic T-invariant only corresponds to a part of transitions. Therefore, if a transition sequence corresponding to some basic T-invariant can be fired, then this transition sequence can be fired infinitely, which means that other transitions can never be fired in such an infinite run. For example, we let

$$ M_0 = \{p_{j,1}, \ r_j \mid j \in \mathbb{N}_5^+\} $$

be the initial marking of the Petri net in Fig. 1.12, and firing the transition $t_{1,1}$ at M_0 leads to the marking

$$ M_1 = \{r_1, \ p_{1,2}, \ r_{j+1}, \ p_{j+1,1} \mid j \in \mathbb{N}_4^+\} $$

which means that the first philosopher are ready for eating. At M_1, the transition sequence $t_{2,1}t_{2,2}t_{2,3}$ corresponding to the T-invariant

$$
\begin{array}{ccccccccccccccc}
t_{1,1} & t_{1,2} & t_{1,3} & t_{2,1} & t_{2,2} & t_{2,3} & t_{3,1} & t_{3,2} & t_{3,3} & t_{4,1} & t_{4,2} & t_{4,3} & t_{5,1} & t_{5,2} & t_{5,3} \\
\left[\begin{array}{ccccccccccccccc}
0 & 0 & 0 & 1 & 1 & 1 & 0 & 0 & 0 & 0 & 0 & 0 & 0 & 0 & 0
\end{array}\right]
\end{array}
$$

can be fired, and thus it can infinitely be fired at M_1. In this case, therefore, the first philosopher can never eat though s/he has been read for eating.

Avoiding Starvation

How to solve this starvation problem?

A starvation-free policy proposed in [7] is that some *dinning cards* are set up for philosophers. Only when a philosopher holds a dinning card, s/he is permitted to change her/his state from *thinking* to *ready for eating*. When s/he holds a dinning card but does not plan to eat, then s/he transfers this dinning card to her/his next neighbour. After s/he gets both chopsticks, s/he also immediately transfers the dinning card to the next neighbour.

Figure 2.4 illustrates this starvation-free solution. A token in the place c_j ($j \in \mathbb{N}_5^+$) means a dining card is in the hand of the jth philosopher. If s/he is not ready for eating, this card is transferred to her/his right neighbour, modelling by the transition $t_{j,4}$. We can compute its basic T-invariants as follows:

$t_{1,1}$	$t_{1,2}$	$t_{1,3}$	$t_{1,4}$	$t_{2,1}$	$t_{2,2}$	$t_{2,3}$	$t_{2,4}$	$t_{3,1}$	$t_{3,2}$	$t_{3,3}$	$t_{3,4}$	$t_{4,1}$	$t_{4,2}$	$t_{4,3}$	$t_{4,4}$	$t_{5,1}$	$t_{5,2}$	$t_{5,3}$	$t_{5,4}$
0	0	0	1	0	0	0	1	0	0	0	1	0	0	0	1	0	0	0	1
1	1	1	0	0	0	0	1	0	0	0	1	0	0	0	1	0	0	0	1
0	0	0	1	1	1	1	0	0	0	0	1	0	0	0	1	0	0	0	1
0	0	0	1	0	0	0	1	1	1	1	0	0	0	0	1	0	0	0	1
0	0	0	1	0	0	0	1	0	0	0	1	1	1	1	0	0	0	0	1
0	0	0	1	0	0	0	1	0	0	0	1	0	0	0	1	1	1	1	0
1	1	1	0	1	1	1	0	0	0	0	1	0	0	0	1	0	0	0	1
1	1	1	0	0	0	0	1	1	1	1	0	0	0	0	1	0	0	0	1
1	1	1	0	0	0	0	1	0	0	0	1	1	1	1	0	0	0	0	1
1	1	1	0	0	0	0	1	0	0	0	1	0	0	0	1	1	1	1	0
0	0	0	1	1	1	1	0	1	1	1	0	0	0	0	1	0	0	0	1
0	0	0	1	1	1	1	0	0	0	0	0	1	1	1	1	0	0	0	1
0	0	0	1	1	1	1	0	0	0	0	1	0	0	0	1	1	1	1	0
0	0	0	1	0	0	0	1	1	1	1	0	1	1	1	0	0	0	0	1
0	0	0	1	0	0	0	1	1	1	1	0	0	0	0	1	1	1	1	0
0	0	0	1	0	0	0	1	0	0	0	1	1	1	1	0	1	1	1	0
1	1	1	0	1	1	1	0	1	1	1	0	0	0	0	1	0	0	0	1
1	1	1	0	1	1	1	0	0	0	0	1	1	1	1	0	0	0	0	1
1	1	1	0	1	1	1	0	0	0	0	1	0	0	0	1	1	1	1	0
1	1	1	0	0	0	0	1	1	1	1	0	1	1	1	0	0	0	0	1
1	1	1	0	0	0	0	1	1	1	1	0	0	0	0	1	1	1	1	0
1	1	1	0	0	0	0	1	0	0	0	1	1	1	1	0	1	1	1	0
0	0	0	1	1	1	1	0	1	1	1	0	1	1	1	0	0	0	0	1
0	0	0	1	1	1	1	0	1	1	1	0	0	0	0	1	1	1	1	0
0	0	0	1	1	1	1	0	0	0	0	1	1	1	1	0	1	1	1	0
0	0	0	1	0	0	0	1	1	1	1	0	1	1	1	0	1	1	1	0
1	1	1	0	1	1	1	0	1	1	1	0	1	1	1	0	0	0	0	1
1	1	1	0	1	1	1	0	1	1	1	0	0	0	0	1	1	1	1	0
1	1	1	0	1	1	1	0	0	0	0	1	1	1	1	0	1	1	1	0
1	1	1	0	0	0	0	1	1	1	1	0	1	1	1	0	1	1	1	0
0	0	0	1	1	1	1	0	1	1	1	0	1	1	1	0	1	1	1	0

When a philosopher (say the first one) enters the *read-for-eating* state with a dining card, her/his neighbours (say the second philosopher) can eat once. If this neighbour plans to eat again, s/he must obtain a dining card again according to the related T-invariants (as shown by the bold parts in the above matrix). However, s/he can obtain a dining card again only when the first philosopher transfers it to her/him. Obviously, only when the first philosopher enters the *eating* state, s/he can transfer

Fig. 2.4 A starvation-free model of the dining philosophers problem [7]. A token in place c_j means that the dining card is held by the jth philosopher. Firing transition $t_{j,4}$ means that the jth philosopher does not plan to eat and then transfers the dining card to the right neighbour. Firing transition $t_{j,1}$ means that the jth philosopher holds a dining card and is ready for eating

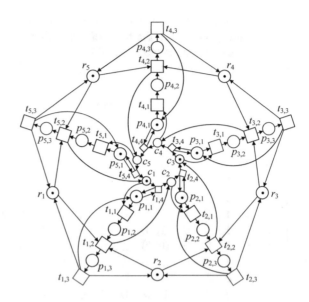

the dining card to the second one. This means that the policy can avoid the starvation states.

Question

Can the number of dining cards be arbitrary to avoid a starvation? Why?

2.6 Summary and Further Reading

These structure concepts of Petri nets can be found in [1, 9] where more properties of them are shown.

The relations discovered in this chapter and these computation algorithms can be found in [4]. Some other methods of computing T-/P-invariants and siphons/traps can be found in [11–15]. However, the problem of computing them is NP-hard theoretically [16]. In the worst case, the number of siphons (respectively traps, minimal T-/P-invariants, and repetitive vectors) of a net grows exponentially in the size of the net.

From this chapter, we see that these structure concepts are closely related to the liveness and deadlock of systems. They can be used to decide the liveness of some systems [1, 9, 17–20] and also to prevent and avoid deadlocks of some systems [21–30]. In the next two chapters, we will show some applications related to these structure concepts.

References

1. Murata, T.: Petri nets: properties, analysis and application. *Proceedings of IEEE* 77: 541–580 (1989)
2. Lautenbach, K., Ridder, H.: Liveness in bounded Petri nets which are covered by T-invariants. In: *the 15th International Conference on Application and Theory of Petri Nets*, pp. 358–375 (1994)
3. Martínez, J ., Silva, M.: A simple and fast algorithm to obtain all invariants of a generalised Petri net. In: *Proceedings of the Second European Workshop on Application and Theory of Petri Nets*, pp. 301–310 (1982)
4. Liu, G.J., Jiang, C.J.: Incidence matrix based methods for computing repetitive vectors and siphons of Petri net. *Journal of Information Science and Engineering* 25: 121–136 (2009)
5. Silva, M., Teruel, E., Colom, J.: Linear algebraic and linear programming techniques for the analysis of place/transition net systems. *Lecture Notes on Computer Sciences* 1491: 309–373 (1996)
6. Liu, G.J., Jiang, C.J., Chao, D.Y.: . A necessary and sufficient condition for the liveness of normal nets. *The Computer journal* 54: 157–163 (2011)
7. Wu, Z.H.: *Introduction to Petri Nets*. China Machine Press, Beijing (2006) (in Chinese)
8. Dickson, L.E.: Finiteness of the odd perfect and primitive abundant numbers with n distinct prime factors. *American Journal of Mathematics* 35: 413–422 (1913)
9. Desel, J., Esparza, J.: *Free Choice Petri Nets*. Cambridge University Press (1995)
10. Yang, S.L., Zhuo, J., Guo, J.: Inverse petri nets: properties and applications. *IFAC Proceedings Volumes* 24: 91–95 (1991)
11. Takano, K., Taoka, S., Yamauchi, M., Watanabe, T.: Experimental evaluation of two algorithms for computing Petri net invariants. *IEICE Transactions on Fundamentals of Electronics, Communications and Computer Sciences* E84-A: 2871–2880 (2001)
12. Tricas, F., Ezpeleta. J.: Computing minimal siphons in Petri net models of resource allocation systems: A parallel solution. *IEEE Transactions on Systems, Man, and Cybernetics – Part A: Systems and Humans* 36: 532-539 (2006)
13. Han, X., Chen, Z., Liu, Z., Zhang, Q.: Calculation of siphons and minimal siphons in Petri nets based on semi-tensor product of matrices. *IEEE Transactions on Systems, Man, and Cybernetics: Systems* 47: 531-536 (2015)
14. Liu, G., Barkaoui, K.: A survey of siphons in Petri nets. *Information Sciences* 363:198–220 (2016)
15. Wang, S., Duo, W., Guo, X., Jiang, X., You, D., Barkaoui, K., Zhou, M.C.: Computation of an emptiable minimal siphon in a subclass of Petri nets using mixed-integer programming. *IEEE/CAA Journal of Automatica Sinica* 8: 219–226 (2020)
16. Yamauchi, M., Watanabe, T.: Time complexity analysis of the minimal siphon extraction problem of Petri nets. *IEICE Transactions on Fundamentals of Electronics, Communications and Computer Sciences* E82-A: 2558–2565 (1999)
17. Liu, G.J., Jiang, C.J., Zhou, M.C.: Process nets with channels. *IEEE Transactions on Systems, Man, and Cybernetics – Part A: Systems and Humans* 42: 213–225 (2012)
18. Liu, G.J., Jiang, C.J., Zhou, M.C., Xiong, P.C.: Interactive Petri nets. *IEEE Transactions on Systems, Man, and Cybernetics: Systems* 43: 291–302 (2013)
19. Liu, G.J., Jiang, C.J., Zhou, M.C., Ohta, A.: The liveness of WS^{3}PR: complexity and decision. *IEICE Transactions on Fundamentals of Electronics, Communications and Computer Sciences* E96-A: 1783–1793 (2013)
20. Liu, G.J., Chen, L.J.: Deciding the liveness for a subclass of weighted Petri nets based on structurally circular wait. *International Journal of Systems and Sciences* 47: 1533–1542 (2016)
21. Ezpeleta, J., Colom, J.M., Martinez, J.: A Petri net based deadlock prevention policy for flexible manufacturing systems. *IEEE transactions on Robotics and Automation* 11: 173–184 (1995)
22. Park, J., Reveliotis, S.A.: Deadlock avoidance in sequential resource allocation systems with multiple resource acquisitions and flexible routings. *IEEE Transactions on Automatic Control* 46: 1572–1583 (2001)

23. Li, Z.W., Zhou, M.C.: Elementary siphons of Petri nets and their application to deadlock prevention in flexible manufacturing systems. *IEEE Transactions on Systems, Man, and Cybernetics – Part A: Systems and Humans* 34: 38–51 (2004)

24. Liu, G.J., Jiang, C.J., Wu, Z.H., Chen, L.J.: A live subclass of petri nets and their application in modelling flexible manufacturing systems. *International Journal of Advanced Manufacturing Technology* 41: 66–74 (2009)

25. Wu, N.Q., Zhou, M.C.: *System Modeling and Control With Resource-Oriented Petri Nets*. New York: CRC Press (2010)

26. Liu, G.J., Jiang, C.J., Zhou, M.C.: Two simple deadlock prevention policies for S^3PR based on key-resource/operation-place pairs. *IEEE Transactions on Automation Science and Engineering* 7: 945–957 (2010)

27. Chao, D., Liu, G.J.: A simple suboptimal siphon-based control model of a well-known S^3PR. *Asian Journal of Control* 14: 163–172 (2012)

28. Fang, H., Liu, G.J., Fang, X.W.: Sufficient and necessary conditions to guarantee deadlock-free scheduling for an extended S^3PR with correlated resources. *Journal of the Chinese Institute of Engineers* 41: 473–482 (2018)

29. Du, N., Hu, H.S., Zhou, M.C.: Robust deadlock avoidance and control of automated manufacturing systems with assembly operations using Petri nets. *IEEE Transactions on Automation Science and Engineering* 17: 1961–1975 (2020)

30. Čapkovič, F.: Modelling and control of resource allocation systems within discrete event systems by means of Petri nets – Part 1: invariants, siphons and traps in deadlock avoidance. *Computing and Informatics* 40: 648–689 (2021)

Chapter 3
Petri Nets with Special Structures

3.1 Single Input and Single Output

State machines and marked graphs are two very simple subclasses of Petri nets, the former requires that each transition has exactly one input and exactly one output, and the latter requires that each place has exactly one input and exactly one output. More details can be found in [1, 2].

3.1.1 State Machine

3.1.1.1 Definition of State Machine

Definition 3.1 (*State Machine*) A net $N = (P, T, F)$ is called a *state machine* if every transition has exactly one input and exactly one output, i.e.,

$$\forall t \in T : |{}^{\bullet}t| = |t^{\bullet}| = 1.$$

Obviously, firing a transition at a marking of a state machine does not change the total number of tokens in this net, and it only moves a token from the input place of this transition into its output place.

Therefore, a state machine with an arbitrary initial marking is bounded, and $[1, 1, \ldots, 1]$ indexed by all places is a P-invariant since each row of its incidence matrix has exactly one 1 and exactly one -1.

© The Author(s), under exclusive license to Springer Nature Singapore Pte Ltd. 2022
G. Liu, *Petri Nets*,
https://doi.org/10.1007/978-981-19-6309-4_3

If a circuit of a state machine is marked at a marking, then firing the transitions of this circuit at this marking in turn will make this marking be reached again.[1]

> Therefore, each circuit of a state machine corresponds to a T-invariant, i.e., every entry of such a T-invariant is the number of times that the corresponding transition occurs in the circuit.

Example 3.1 The state machine in Fig. 3.1a has three elementary circuits:

$$p_3 t_4 p_4 t_5 p_3, \quad p_1 t_1 p_2 t_2 p_5 t_7 p_1, \quad \text{and} \quad p_1 t_1 p_2 t_3 p_3 t_4 p_4 t_6 p_5 t_7 p_1.$$

They exactly correspond to the following three minimal T-invariants

$$[0, 0, 0, 1, 1, 0, 0], \quad [1, 1, 0, 0, 1, 0, 1] \quad \text{and} \quad [1, 1, 1, 1, 0, 1, 1]$$

indexed by $t_1, t_2, \ldots$, and t_7. Note that the first two circuits are minimal, but the last one is not minimal because places p_3 and p_4 form another circuit.

Fig. 3.1 a A state machine in which there is a *choice* at place p_2 and transitions t_4 and t_5 form a *loop*. **b** A marked graph in which t_1 creates two *parallel subprocesses*, there is an *interaction* between t_4 and t_5 of the two subprocesses, and the two subprocesses are *synchronised* by t_6. Note: transitions t_2 and t_4 in **b** are not viewed as a loop structure, but they can be thought of as requiring-releasing a resource/flag

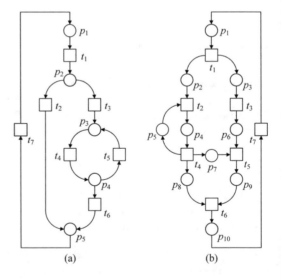

(a)   (b)

---

[1] A sequence $x_1 x_2 \ldots x_k$ is a *circuit* of a net $N = (P, T, F)$ if $k > 1$, $x_1 = x_k$ and $(x_1, x_2), \ldots, (x_{k-1}, x_k) \in F$. A circuit $x_1 x_2 \ldots x_k$ is *elementary* if any two nodes except x_1 and x_k are different in it. Later, we also use the concept of *minimal circuit*: a circuit $x_1 x_2 \ldots x_k$ is *minimal* if the set of places in it does not properly include the set of places in any other circuit [3, 4]. Obviously, a minimal circuit is elementary. A circuit is called a *siphon circuit* (respectively, *trap circuit*) if all places in the circuit form a siphon (respectively, trap).

State machines are a very special subclass of Petri nets so that their many properties can be decided easily.

3.1.1.2 Liveness of State Machine

Theorem 3.1 ([1]) *A connected state machine N with an initial marking M_0 is live if and only if*

1. *N is strongly connected, and*
2. $M_0 \neq \overline{0}$.

Proof (Necessity) If every place of a state machine (N, M_0) is unmarked at the initial marking, then every transition is disabled at the initial marking because every transition has an unmarked input place. Hence, $M_0 \neq \overline{0}$ if (N, M_0) is live. If a connected state machine is not strongly connected, then there exists a place p and a non-empty set S of places such that for every place in S there is a directed path from it to p but not vice versa. In this case, each token in S can be moved into p along with a directed path and finally no token can enter into S, making all transitions that are associated with S disabled and contradicting the fact that (N, M_0) is live. Therefore, a live and connected state machine must be strongly connected.

(Sufficiency) If a state machine N is strongly connected and the initial marking M_0 is not equal to $\overline{0}$, then for each reachable marking $M \in R(N, M_0)$ and each transition t, we can move a token from a place p marked at M to the input place of t along with a directed path from p to t and thus enable t. Therefore, a strongly connected state machine with a non-empty initial marking is live. □

Based on Theorem 3.1, the decision of the liveness problem of a state machine with a non-empty initial marking is transferred into the decision of the strong-connectivity of a directed graph, which can be done in polynomial time [6, 7]:

$$\mathcal{O}(|P| + |T| + |F|) \xlongequal{|F|=2|T|} \mathcal{O}(|P| + 3|T|).$$

3.1.1.3 Reachability of State Machine

Theorem 3.2 ([1]) *Let the state machine $N = (P, T, F)$ be strongly connected and M_0 be an initial marking. A marking M is reachable in (N, M_0) if and only if*

$$M(P) = M_0(P).$$

Proof (Necessity) Since $[1, 1, \ldots, 1]$ is a P-invariant, we have $M(P) = M_0(P)$ for each $M \in R(N, M_0)$ according to the characteristics of P-invariants.

(Sufficiency) In what follows, the sufficiency is shown from two aspects.

- When $M(P) = M_0(P) = 0$, the conclusion holds obviously.
- When $M(P) = M_0(P) > 0$, we can make $M_0(P)$ copies of N and each copy contains only one token in its initial marking. In other words, for each place p, if $M_0(p) > 0$ in (N, M_0), then we make $M_0(p)$ copies of N and put only one token into the place p of each copy (thereby forming its initial marking). Next, from the $M_0(P)$ copies, we select $M(p)$ ones if $M(p) > 0$ for some place p; and for each one of the $M(p)$ selected copies we can move its initial token into place p due to its strong connectivity. Since these copies are independent to each other and live (according to Theorem 3.1), it is easy to understand that the marking M can be reached from M_0 in N when the $M_0(P)$ copies are thought of as a whole.

$\square$

> Theorem 3.2 means that the time complexity of the reachability problem of live state machines is polynomial: $\mathcal{O}(|P|)$.

From the definition of state machine and the above example, it is easy to see that the structures of a state machine can represent *sequence*, *choice* and *loop* of actions of a process but cannot generally represent *parallel*, *synchronisation* and *interaction* of actions. However, the latter can be represented in a marked graph.

3.1.2 Marked Graph

3.1.2.1 Definition of Marked Graph

Definition 3.2 (*Marked Graph*) A net $N = (P, T, F)$ is called a *marked graph* if every place has exactly one input and exactly one output, i.e.,

$$\forall p \in P : |{}^{\bullet}p| = |p^{\bullet}| = 1.$$

The first feature observed is that firing any transition in an elementary circuit of a marked graph cannot change the total number of tokens in this circuit. Therefore, any elementary circuit of a marked graph corresponds to a P-invariant. More generally, any circuit of a marked graph corresponds to a P-invariant since each circuit can be decomposed into a group of elementary circuits, i.e., every entry of the P-invariant is the number of times that the corresponding place occurs in the circuit.

Example 3.2 The marked graph in Fig. 3.1b has four elementary circuits

$$p_5 t_2 p_4 t_4 p_5, \quad p_1 t_1 p_2 t_2 p_4 t_4 p_8 t_6 p_{10} t_7 p_1,$$

$$p_1 t_1 p_3 t_3 p_6 t_5 p_9 t_6 p_{10} t_7 p_1, \quad p_1 t_1 p_2 t_2 p_4 t_4 p_7 t_5 p_9 t_6 p_{10} t_7 p_1.$$

They correspond to the following minimal P-invariants:

$$[0, 0, 0, 1, 1, 0, 0, 0, 0, 0], \quad [1, 1, 0, 1, 0, 0, 0, 1, 0, 1],$$

$$[1, 0, 1, 0, 0, 1, 0, 0, 1, 1], \quad [1, 1, 0, 1, 0, 0, 1, 0, 1, 1].$$

Here, the four P-invariants are all indexed by the places $p_1, p_2, \ldots,$ and p_{10}.

Given an arbitrary state machine, it is always bounded for any initial marking; however, marked graphs are not of such a property. For instance of the Petri net in Fig. 1.2b, if we add an output transition for the right place, then we obtain a marked graph which is obviously unbounded. Fortunately, if a marked graph is strongly connected, it is bounded for any initial marking since all places form a P-invariant due to the strong connectivity.

3.1.2.2 Liveness of Marked Graph

Theorem 3.3 ([1]) *A marked graph with an initial marking is live if and only if every (elementary) circuit has a token at the initial marking.*

Proof (Necessity) Since each place of a marked graph has exactly one input and exactly one output, the places in a circuit in a marked graph form both a siphon and a trap. Therefore, all transitions in a circuit are disabled at some marking if the circuit has no token at this marking. Therefore, if a marked graph with an initial marking is live, then every (elementary) circuit has a token at this initial marking.

(Sufficiency) Since the places in a circuit in a marked graph form both a siphon and a trap, a circuit will be marked at any reachable marking if it is marked at the initial marking. The next section will show that marked graphs are a subclass of free-choice nets and a free-choice net with an initial marking is live if and only if each siphon contains an initially-marked trap. Therefore, if every (elementary) circuit of a marked graph is marked at an initial marking, then it is live at the initial marking. $\square$

Obviously, the liveness decision of a marked graph with an initial marking can be transferred into the generation of all elementary circuits of a directed graph, while this computation can be conducted in polynomial time [8, 9]:

$$\mathcal{O}((|P| + |T| + |F|) \cdot (k + 1)) \xrightarrow{|F|=2|P|} \mathcal{O}((3|P| + |T|) \cdot (k + 1)),$$

where k is the number of elementary circuits.

Fortunately, a more efficient algorithm can be provided: we first delete all initially-marked paces and those arcs connected with them, and then go to decide if the resulting net has a circuit. To perform such a decision, we only check those transitions that has no input or output. If a transition has no input, then it and its output places are continually deleted. The above operations are continually conducted until the net becomes empty or the rest transitions cannot be deleted. The case that the final net becomes empty indicates that every circuit in the original net is marked, and another case means that there is a circuit unmarked at the initial marking. Obviously, this decision procedure can be done in $\mathcal{O}(|P| + |T| + |F|)$.

3.1.2.3   Reachability of Marked Graph

Theorem 3.4 ([2]) *Let the marked graph $N = (P, T, F)$ with an initial marking M_0 be live. A marking M is reachable in (N, M_0) if and only if the state equation*

$$M = M_0 + X \cdot [N]$$

has a nonnegative integer solution.

To prove the sufficiency of Theorem 3.4, we need the following lemmata.

Lemma 3.1 ([2]) *Let N be a marked graph, M be a marking and X be a nonnegative integer vector indexed by transitions of N. Then, $M(p) > 0$ for each place $p \in {}^\bullet\|X\| \setminus \|X\|^\bullet$ if the following holds:*

$$M + X \cdot [N] \geq \overline{0},$$

where $\|X\| = \{t \mid X(t) > 0\}$.

Proof Because each column of $[N]$ has only one -1 and only one 1, we have that for each place $p \in {}^\bullet\|X\| \setminus \|X\|^\bullet$, a transition as the output of p belongs to $\|X\|$, but another transition as its input does not belong to $\|X\|$. Hence, for each place $p \in {}^\bullet\|X\| \setminus \|X\|^\bullet$, we have
$$X \cdot [N](p) \leq -1.$$

Thus, we have $M(p) > 0$ due to $M(p) + X \cdot [N](p) \geq 0$.   □

Lemma 3.2 ([2]) *Let the marked graph N with an initial marking M_0 be live and X be a nonnegative integer vector indexed by transitions of N. If the following holds:*

$$M_0 + X \cdot [N] \geq \overline{0} \wedge X \neq \overline{0},$$

then there exists a transition $t \in \|X\|$ such that

$$M_0[t\rangle.$$

Fig. 3.2 The illustration of the proof of Lemma 3.2

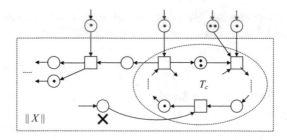

Proof According to Lemma 3.1 we know that each place in $^\bullet||X|| \setminus ||X||^\bullet$ is marked at M_0. Now we consider transitions in $||X||$ from the following two cases:

- $\exists t \in ||X||$: $^\bullet t \subseteq {}^\bullet||X|| \setminus ||X||^\bullet$. Obviously, such a transition is enabled at M_0.
- $\forall t \in ||X||$: $^\bullet t \not\subseteq {}^\bullet||X|| \setminus ||X||^\bullet$. For this case, there must exist a non-empty $T_c \subseteq ||X||$ satisfying that all nodes in $T_c \cup ({}^\bullet T_c \setminus ({}^\bullet||X|| \setminus ||X||^\bullet))$ are pairwise strongly connected (or else the first case occurs), i.e., these nodes form a strongly connected subnet as shown in Fig. 3.2. Since (N, M_0) is live, such a subnet is also live at

$$M_0 \upharpoonright ({}^\bullet T_c \setminus ({}^\bullet||X|| \setminus ||X||^\bullet)).$$

Therefore, there is a transition in T_c whose input places in this subnet are all marked at

$$M_0 \upharpoonright ({}^\bullet T_c \setminus ({}^\bullet||X|| \setminus ||X||^\bullet)).$$

Therefore, all input places of such a transition are marked at M_0 even when some of these input places are in $^\bullet||X|| \setminus ||X||^\bullet$.

$\square$

According to Lemma 3.2, firing such a transition t at M_0 leads to the following results:

- $M_0 + X \cdot [N] = M_0 + \psi(t) \cdot [N] + (X - \psi(t)) \cdot [N] \geq \bar{0}$,
- $X - \psi(t)$ is still a nonnegative integer vector, and
- $(N, M_0 + \psi(t) \cdot [N])$ is still live.

In other words, we can still find a transition $t' \in ||X - \psi(t)||$ that is enabled at the marking

$$M_0 + \psi(t) \cdot [N]$$

if

$$X - \psi(t) \neq \bar{0}.$$

In consequence, we can finally find a firable transition sequence σ in (N, M_0) such that

$$\psi(\sigma) = X.$$

Lemma 3.3 ([2]) *Let the marked graph N with an initial marking M_0 be live and X be a nonnegative integer vector satisfying*

$$M_0 + X \cdot [N] \geq \overline{0}.$$

Then, there is a firable transition sequence σ in (N, M_0) such that

$$\psi(\sigma) = X.$$

Based on Lemma 3.3 we know that the sufficiency of Theorem 3.4 holds.

> Although deciding the reachability problem of live marked graphs can be transferred into solving an integer programming problem, the latter is not too easy since it is harder compared with the linear programming problem [2].

Question

Try to explore the complexity of the reachability problem of marked graphs.

3.2 Choice Structures

Two kinds of choice structures, i.e., free choice and asymmetric choice, are introduced in this section. The former considers symmetry while the latter allows asymmetry. More details can be found in [1, 2].

3.2.1 Free-Choice Net

3.2.1.1 Definition of Free-Choice Net

A free-choice net requires that any two places either have the same outputs or have no any common output. According to symmetry, this property means that any two transitions either have the same inputs or have no any common input. Here a simplified definition is given, but a net satisfying the above requirement can be transformed into such a simplified version.

Definition 3.3 (*Free-Choice Net*) A net $N = (P, T, F)$ is called a *free-choice net* if for any two transitions $t_1 \in T$ and $t_2 \in T$ we have

Fig. 3.3 a Free-choice
structure; and **b** extended
free-choice structure that can
be transferred into the
free-choice form

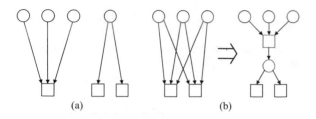

(a)   (b)

$$(^\bullet t_1 \cap {}^\bullet t_2 \neq \emptyset \wedge t_1 \neq t_2) \Rightarrow |^\bullet t_1| = |^\bullet t_2| = 1.$$

A free-choice net with an initial marking is called a *free-choice Petri net.*

Figure 3.3a shows the free-choice structures. Obviously, a state machine and a marked graph are both a free-choice net. There is a more general definition of free choice that is called *extended free-choice* and requires that

$$^\bullet t_1 \cap {}^\bullet t_2 \neq \emptyset \Rightarrow {}^\bullet t_1 = {}^\bullet t_2.$$

It has been shown that an extended free-choice Petri net can be transferred into a free-choice Petri net while preserving its behaviours [10], as shown in Fig. 3.3b.

Question

Try to provide a proof about the behaviour-preserving problem when transferring an extended free-choice Petri net into a free-choice one.

3.2.1.2   Liveness of Free-Choice Net

The liveness of a free-choice Petri net is closely related to its siphons and traps.

Theorem 3.5 ([1]) *A free-choice Petri net is live if and only if each siphon includes a trap marked at the initial marking.*

When a trap is marked at a marking, it will be marked at each reachable marking. Therefore, if each siphon includes a trap marked at the initial marking, then they will be marked at each reachable marking. Therefore, the following conclusion can directly derive the sufficiency of Theorem 3.5.

Lemma 3.4 ([1]) *If a free-choice Petri net is not live, then there must be a siphon that becomes unmarked at some reachable marking.*

Proof When a free-choice Petri net (N, M_0) is not live, there must be a reachable marking and a transition, say $M \in R(N, M_0)$ and t, such that

$$\forall M' \in R(N, M) : \neg M'[t\rangle.$$

According to the characteristic of free-choice nets, there is a place $p \in {}^\bullet t$ such that

$$\forall M' \in R(N, M) : M'(p) = 0.$$

The reason can be explained from the following two cases:

– if $|{}^\bullet t| = 1$, i.e., the place p is the unique input of t, then we obviously have

$$\forall M' \in R(N, M) : M'({}^\bullet t) = 0.$$

– if $|{}^\bullet t| > 1$, then each input place of t has exactly one output transition, namely t. Therefore, when some input place of t is marked, it will be marked at any succeeding marking since t is disabled and thus the token(s) in this place will not be removed forever. Therefore, there must be a place $p \in {}^\bullet t$ unmarked at each $M' \in R(N, M)$ since t is disabled at each M'.[2]

Consequently, each input transition of p is disabled at each reachable marking $M' \in R(N, M)$. Similarly, for such a transition, there must also be an input place that is unmarked at each marking $M' \in R(N, M)$. Due to the finiteness of the number of places and transitions, we finally can find a set of places P' such that

$$ {}^\bullet P' \subseteq P'^\bullet,$$

i.e., they form a siphon and are unmarked at M.   □

The necessity of Theorem 3.5 can be ensured by the following lemma:

Lemma 3.5 ([1]) *If a free-choice Petri net is live, then each siphon includes an initially-marked trap.*

Proof Obviously, if a siphon is a trap, then it must be marked at the initial marking (or else all transitions related to the siphon are disabled at each reachable marking). Now, we consider the case that a siphon is not a trap and let $P_0 \subseteq P$ be such a siphon

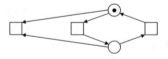

Fig. 3.4 For a dead transition, e.g. the most left one, its input places are marked alternatively in the run but unmarked simultaneously

---

[2] This case can occur in free-choice Petri nets as well as asymmetric-choice Petri nets, but does not necessarily occur in more general Petri nets, because for a dead transition in a general Petri net, its input places can be marked alternatively in the run but cannot be marked simultaneously, as shown in Fig. 3.4.

in the free-choice Petri net $(N, M_0) = (P, T, F, M_0)$. At this time, there is at least one transition that belongs to $P_0^\bullet$ but not to $^\bullet P_0$, i.e., such a transition can remove tokens out of this siphon but not create a token into it. We can define such a subnet associated with places in P_0 except those ones through which tokens can flow out of P_0:

$- T_1 = P_0^\bullet \setminus {}^\bullet P_0,$
$- P_1 = P_0 \setminus \{p \in P_0 \mid p^\bullet \cap T_1 \neq \emptyset\},$
$- F_1 = F \cap ((P_1 \times {}^\bullet P_0) \cup ({}^\bullet P_0 \times P_1)),$
$- N_1 = (P_1, {}^\bullet P_1 \cup P_1^\bullet, F_1).$

First, we know that P_1 must have tokens at M_0; or else (i.e., if they are all in $P_0 \setminus P_1$), firing transitions in T_1 in (N, M_0) can empty siphon P_1 and thus makes (N, M_0) not live. Additionally, $(N_1, M_0 \upharpoonright P_1)$ is also live; or else (i.e., if it is not live), there is a siphon in the free-choice net N_1 such that it is not marked at $M_0 \upharpoonright P_1$, and thus this siphon (that is also a siphon in N) is not marked at M_0, which means that (N, M_0) is not live. Now, we consider $(N_1, M_0 \upharpoonright P_1)$. If $P_1^\bullet \subseteq {}^\bullet P_1$, then that is exactly what we want. If $P_1^\bullet \nsubseteq {}^\bullet P_1$, we can define another subnet associated with places in P_1 except those ones through which tokens can flow out of P_1:

$- T_2 = P_1^\bullet \setminus {}^\bullet P_1,$
$- P_2 = P_1 \setminus \{p \in P_1 \mid p^\bullet \cap T_2 \neq \emptyset\},$
$- F_2 = F_1 \cap ((P_2 \times {}^\bullet P_1) \cup ({}^\bullet P_1 \times P_2)) = F \cap ((P_2 \times {}^\bullet P_1) \cup ({}^\bullet P_1 \times P_2)),$
$- N_2 = (P_2, {}^\bullet P_2 \cup P_2^\bullet, F_2).$

Similarly, P_2 must have tokens at M_0; or else (i.e., if they are all in $P_1 \setminus P_2$), firing transitions in T_2 in (N, M_0) can move them from $P_1 \setminus P_2$ into $P_0 \setminus P_1$, and then firing transitions in T_1 empties siphon P_1. Similarly, $(N_2, M_0 \upharpoonright P_2)$ is also live. Therefore, if $P_2^\bullet \subseteq {}^\bullet P_2$, that is just what we want; or else we can continue to perform the above operations. When such a subnet, namely $N_k = (P_k, {}^\bullet P_k \cup P_k^\bullet, F_k)$, is produced every time, it must have three constraints: $P_k \neq \emptyset$, P_k is marked at M_0, and $(N_k, M_0 \upharpoonright P_k)$ is live. Therefore, there must be such a subnet such that $P_k^\bullet \subseteq {}^\bullet P_k$, or else we have to infinitely produce them with the above three constraints, that is however impossible due to the finiteness of the number of places. □

Example 3.3 The top in Fig. 3.5 shows a free-choice net that is live at the shown marking. All places form a siphon but it is not a trap due to the existence of the most left and most right transitions. Then deleting the most left and most right places yields the subnet as shown in the middle part in Fig. 3.5, but the places in the subnet does not form a trap since the place holding two tokens has an output transition that is not the input of this subnet. Therefore, we can continue to delete this place and finally obtain the subnet as shown at the bottom in Fig. 3.5. Obviously, the final two places form a trap initially marked.

Fig. 3.5 Illustration of the proof of the necessity of Theorem 3.5. The example comes from [11]. The parts in the middle and the bottom show the deleted places in turns, and the remaining in the bottom is a trap

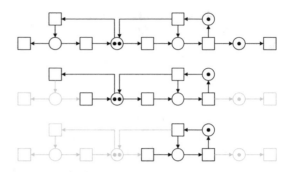

Although there is such a beautiful condition to decide the liveness for free-choice Petri nets, the decision complexity is still very high in theory: NP-complete [1, 12]. Fortunately, the liveness of a free-choice Petri net can be decided in polynomial time if the Petri net is safe [1]. So far such a similar condition of deciding the reachability problem of free-choice Petri nets has not been given. The decision complexity of this reachability problem is EXPSPACE-hard in theory [1], and it is still PSPACE-complete for safe free-choice Petri nets [1].

Question

Try to prove the above conclusions of complexity. Try to provide an polynomial-time algorithm of deciding the liveness of safe free-choice Petri nets. Try to analyse the complexity of deciding if a free-choice Petri net is safe or not.

3.2.2   Asymmetric-Choice Net

3.2.2.1   Definition of Asymmetric-Choice Net

An asymmetric-choice net requires that when two places have a common output, all outputs of one of the two places must be the outputs of another one.

Definition 3.4 (*Asymmetric-Choice Net*) A net $N = (P, T, F)$ is called an *asymmetric-choice net* if for any two places $p_1 \in P$ and $p_2 \in P$ we have

$$p_1^\bullet \cap p_2^\bullet \neq \emptyset \Rightarrow (p_1^\bullet \subseteq p_2^\bullet \vee p_2^\bullet \subseteq p_1^\bullet).$$

An asymmetric-choice net with an initial marking is called an *asymmetric-choice Petri net*.

Question

Try to design a polynomial-time algorithm of deciding if a net is an asymmetric-choice net or not.

According to Definition 3.4, it is easy to derive that if places p_1, p_2, ..., and p_k have a common output transition in an asymmetric-choice net, then there is an inclusion relation among $p_1^{\bullet}$, $p_2^{\bullet}$, ..., and $p_k^{\bullet}$ [1], i.e.,

there is a rearrangement of 1, 2, ..., and k (say j_1, j_2, ..., and j_k) such that

$$p_{j_1}^{\bullet} \subseteq p_{j_2}^{\bullet} \subseteq \cdots \subseteq p_{j_k}^{\bullet}.$$

This property will be used in the next chapter.

Asymmetric-choice Petri nets can well describe the control flow of the interaction/collaboration of multiple processes. Multiple processes interact and collaborate via sending and receiving messages, and a process can choose different runs according to the different messages from other processes, as shown in Fig. 3.6a. Obviously, a free-choice net is a special asymmetric-choice net.

3.2.2.2 Liveness of Asymmetric-Choice Net

The liveness of an asymmetric-choice Petri net is also closely related to its siphons.

Theorem 3.6 ([1]) *An asymmetric-choice Petri net is live if and only if each siphon is marked at each reachable marking.*

Proof (Sufficiency) The conclusion in Lemma 3.4 and its proof are also available for asymmetric-choice Petri nets, i.e., if an asymmetric-choice Petri net is not live, then there is a siphon and a reachable marking such that this siphon is unmarked at this reachable marking. It is based on such a characteristic: when a transition in an asymmetric-choice Petri net is dead at a marking, there is an input place of this transition that is unmarked forever in the future. Therefore, the sufficiency of Theorem 3.6 holds.

(Necessity) The necessity of Theorem 3.6 holds obviously since such a necessary condition works for any Petri net. □

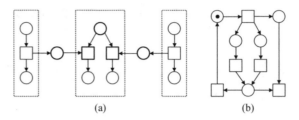

(a)   (b)

Fig. 3.6 **a** The asymmetric-choice structure can model the control flow of the interaction and collaboration of multiple processes. **b** A live asymmetric-choice Petri net in which every siphon does not contain any trap

The decision condition of the liveness of asymmetric-choice Petri nets in Theorem 3.6 is not as perfect as that of free-choice Petri nets in Theorem 3.5 since the latter is only dependent on the net structures and initial marking.

It has been theoretically proven that the complexity of solving the liveness problem of safe asymmetric-choice Petri nets is PSPACE-complete [13], while the complexity of solving the liveness problem of safe free-choice Petri nets is polynomial-time [1]. Certainly, if each siphon contains an initially marked trap in an asymmetric-choice Petri net, then it is live; but not vice versa as shown in Fig. 3.6b.

Question

Try to prove the above conclusion of complexity.

3.3   Circuit Structure

Just as shown in the above sections, one important reason of hard deciding liveness and reachability lies in the existence of circuits, and the existence of traps is usually a sufficient condition leading to a live Petri net. This section introduces two subclasses of nets whose structures are closely related to circuits and traps. For more details, one can read [2–5].

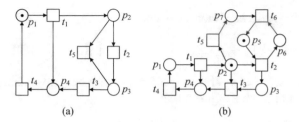

(a)   (b)

Fig. 3.7 **a** A normal Petri net that has only one minimal circuit $p_1 t_1 p_4 t_4 p_1$ marked initially. **b** A normal Petri net that has two minimal circuits $p_1 t_1 p_4 t_4 p_1$ and $p_5 t_2 p_6 t_6 p_5$ in which the former is initially unmarked

3.3.1 Normal Net

3.3.1.1 Definition of Normal Net

A normal net requires that each minimal circuit is a trap circuit. The definition of minimal circuit can be found in Sect. 3.1.

Definition 3.5 (*Normal Net*) A net $N = (P, T, F)$ is called a *normal net* if the places in each minimal circuit in N form a trap. A normal net with an initial marking is called a *normal Petri net*.

Example 3.4 The marked graph in Fig. 3.1b is a normal net but the state machine in Fig. 3.1a is not. In fact, every marked graph is a normal net since every circuit in a marked graph is a trap circuit. Figure 3.7a, b also show two normal nets.

In a normal Petri net, any marked minimal circuit cannot become empty and the total number of tokens in each minimal circuit does not decrease in the runs [3].

Question

Try to prove the above conclusions. And try to develop an algorithm of deciding if a net is normal or not and analyse its complexity.

3.3.1.2 Reachability of Normal Net

These special structures and characteristics ensure that the set of reachable markings of a normal Petri net is *semi-linear* [3].

A set $\mathcal{L} \subseteq \mathbb{N}^m$ is *linear* if there are elements $X_0, X_1, \ldots, X_k \in \mathbb{N}^m$ such that

$$\mathcal{L} = \{X_0 + j_1 \cdot X_1 + \cdots + j_k \cdot X_k \mid j_1, \ldots, j_k \in \mathbb{N}\}$$

where $\mathbb{N}^m$ is the set of all m-dimensional nonnegative integer vectors. A set is *semi-linear* if it is a finite union of linear sets.

Example 3.5 For instance, the reachable marking set of the normal Petri net in Fig. 3.7a can be represented as

$$\{X_1 + j_1 \cdot X_1 + j_2 \cdot X_2 + j_3 \cdot X_3 + j_4 \cdot X_4 \mid j_1, j_2, j_3, j_4 \in \mathbb{N}\}$$
$$\cup \{X_4 + j_1 \cdot X_1 + j_2 \cdot X_2 + j_3 \cdot X_3 + j_4 \cdot X_4 \mid j_1, j_2, j_3, j_4 \in \mathbb{N}\},$$

where $X_1 = [1, 0, 0, 0], X_2 = [0, 1, 0, 0], X_3 = [0, 0, 1, 0]$ and $X_4 = [0, 0, 0, 1]$ are four markings.

The reachable marking set of a normal Petri net is semi-linear [3]. The knowledge of semi-linear sets is beyond the scope of this book and thus not be introduced any more. For a normal Petri net in which each minimal circuit is initially marked, deciding its reachability can be transformed into solving its state equation.

Theorem 3.7 ([3]) *Let $(N, M_0) = (P, T, F, M_0)$ be a normal Petri net such that every minimal circuit is marked at M_0. A marking M is reachable in (N, M_0) if and only if there is a nonnegative integer vector X such that*

$$M = M_0 + X \cdot [N].$$

The necessity holds for any Petri net, while the sufficiency is based on the following fact:

Lemma 3.6 ([3]) *Let $N = (P, T, F)$ be a net, M be a marking, and X be a nonnegative integer vector such that*

$$M + X \cdot [N] \geq \overline{0}.$$

If any transition t with $X(t) > 0$ is disabled at M, then there exists a circuit that is unmarked at M.

Proof We let t_1 be a transition such that $X(t_1) > 0$. Because t_1 is disabled at M, there is an input place p such that

$$[N]^-(t_1, p) = 1 \text{ and } M(p) = 0.$$

Due to

$$(M + X \cdot [N])(p) = (X \cdot [N])(p) = \sum_{t \in T} (X(t) \cdot ([N]^+(t, p) - [N]^-(t, p))) \geq 0,$$

there exists a transition t_2 such that

$$[N](t_2, p) = 1 \wedge X(t_2) > 0.$$

Since t_2 is disabled either at M, we know that it also has an input place unmarked at M, and so on. Consequently, we can construct a token-free circuit such that the transitions in the circuit come from $\{t \in T \mid X(t) > 0\}$. $\square$

We are now back to the proof of the sufficiency of Theorem 3.7 as follows.

Proof (Sufficiency of Theorem 3.7) According to Lemma 3.6, there exists a transition in (N, M_0), say t_1, such that

$$M_0[t_1\rangle M_1 \wedge X(t_1) > 0.$$

Since every minimal circuit is still marked at M_1 and the nonnegative integer vector $X_1 = X - \psi(t)$ still satisfies

$$M = M_0 + X \cdot [N] = M_1 + X_1 \cdot [N],$$

there still exists a transition in (N, M_1), say t_2, such that

$$M_1[t_2\rangle M_2 \wedge X_1(t_2) > 0.$$

Consequently, we can find a firable transition sequence $\sigma = t_1 t_2 \ldots$ such that

$$M_0[\sigma\rangle M \wedge \psi(\sigma) = X.$$

Hence, the sufficiency of Theorem 3.7 holds. $\square$

If a normal Petri net has a minimal circuit initially unmarked, then Theorem 3.7 does not hold.

In other words, for such a normal Petri net that has a minimal circuit initially unmarked, when $M = M_0 + X \cdot [N]$ holds for some nonnegative integer vector X, M is not necessarily reachable in (N, M_0). Such an example is shown here.

Example 3.6 Figure 3.7b shows a normal Petri net in which there are only two transitions t_2 and t_5 that are enabled at the initial marking $M_0 = [0, 1, 0, 0, 1, 0, 0] =$

$\{p_2, p_5\}$ and firing each of them will immediately result in a deadlock. Obviously, the marking $M = [1, 1, 0, 0, 1, 0, 0] = \{p_1, p_2, p_5\}$ is not reachable from M_0, but there is $X = [3, 1, 1, 4, 1, 1]$ such that $M = M_0 + X \cdot [N]$. This is because it has an initially unmarked circuit $p_1 t_1 p_4 t_4 p_1$.

The reachability problem of normal Petri nets is NP-complete [14].

Question

Try to provide a proof of this conclusion of complexity.

3.3.1.3   Liveness of Normal Net

Theorem 3.8 ([15]) *A normal Petri net is live if and only if every siphon is marked at every reachable marking.*

The necessity holds for any live Petri net. To prove the sufficiency, we need the following conclusion:

Lemma 3.7 ([15]) *If a normal Petri net satisfies that every place has at least one input and every minimal circuit is initially marked, then it is live.*

Proof First, since a minimal circuit is a trap circuit of a normal Petri net (N, M_0), it will be marked at every reachable marking when it is initially marked. Since every place has at least one input transition in the normal Petri net, there is a positive integer vector X such that[3]

$$X \cdot [N] \geq \overline{0}.$$

Therefore, according to the proof of Theorem 3.7 we know that for each reachable marking $M \in R(N, M_0)$, there is a firable transition sequence σ at M such that

$$\psi(\sigma) = X.$$

In other words, for each transition t and each reachable marking $M \in R(N, M_0)$, there exists a reachable marking $M' \in R(N, M)$ such that

$$M'[t\rangle,$$

---

[3] This property of normal Petri nets without source place and its proof can be found in [15] and omitted here for simplicity. It is also means that a normal net is structurally repetitive if and only if every place has at least one input. Note that such a conclusion does not necessarily hold for general Petri nets and Fig. 3.4 shows such a counterexample.

i.e., the normal Petri net is live. □

To prove the sufficiency of Theorem 3.8, we also need the following conclusion provided by Barkaoui et al. [16]:

Lemma 3.8 ([16]) *For a non-live Petri net, its transition set can be partitioned into two subsets and there exists a reachable marking at which the transitions in one subset are live but each transition in another subset is disabled forever.*

Now, we go back to the proof of the sufficiency of Theorem 3.8.

Proof (Sufficiency of Theorem 3.8) We assume that a normal Petri net $(N, M_0) = (P, T, F, M_0)$ is not live when every place of it has at least one input transition and every siphon is marked at every reachable marking. Note that since every siphon is marked at every reachable marking, there always exists at least one enabled transition at every reachable marking.

Based on Lemma 3.8 and the above assumption we can derive that there exist $T_1 \subset T$, $T_2 \subset T$ and $M \in R(N, M_0)$ such that

- $T = T_1 \cup T_2 \wedge T_1 \cap T_2 = \emptyset \wedge T_1 \neq \emptyset \wedge T_2 \neq \emptyset$, and
- each transition in T_1 is live in (N, M) but each transition in T_2 is dead (i.e., disabled forever) in (N, M).

We define a set $P_0 = \{p \in P \setminus ({}^{\bullet}T_1 \cup T_1^{\bullet}) \mid M(p) = 0\}$ as well as a function $f(t) = {}^{\bullet}t \cap P_0$ for each $t \in T_2$. Obviously, we have

$$P_0^{\bullet} \subseteq T_2 \wedge {}^{\bullet}P_0 \subseteq T_2.$$

Based on the two definitions we can draw a conclusion that there exists at least one transition $t \in T_2$ such that $f(t) = \emptyset$. (By contradiction) If $f(t) \neq \emptyset$ for all $t \in T_2$, then we have

$$P_0^{\bullet} = T_2,$$

i.e., P_0 is a siphon that is empty at M, which results in a contradiction with the fact that each siphon is marked at M.

In consequence, we denote t_0 as such a transition in T_2 that satisfies $f(t_0) = \emptyset$. This means that those input places of t_0, belonging to $P \setminus ({}^{\bullet}T_1 \cup T_1^{\bullet})$, are marked at M (if they exist); in other words, the reason making t_0 dead in (N, M) is that those input places of t_0, belonging to ${}^{\bullet}T_1 \cup T_1^{\bullet}$, cannot be simultaneously marked at every reachable marking of (N, M). Additionally, we denote N_1 as the subnet generated by T_1, i.e.,

$$N_1 = ({}^{\bullet}T_1 \cup T_1^{\bullet}, T_1, F \cap (((T_1 \times ({}^{\bullet}T_1 \cup T_1^{\bullet})) \cup (({}^{\bullet}T_1 \cup T_1^{\bullet}) \times T_1))).$$

Obviously, $(N_1, M \upharpoonright ({}^{\bullet}T_1 \cup T_1^{\bullet}))$ is still a normal Petri net and is live. Furthermore, N_2 is generated based on $T_1 \cup \{t_0\}$ and $({}^{\bullet}T_1 \cup T_1^{\bullet})$, i.e., we add into N_1 transition t_0 and construct the arcs between t_0 and every place in ${}^{\bullet}T_1 \cup T_1^{\bullet}$ in N. Obviously,

N_2 is still a normal net and t_0 is dead in $(N_2, M \upharpoonright ({}^\bullet T_1 \cup T_1^\bullet))$. Since N_2 has no source place, there is a positive integer vector X such that

$$X \cdot [N] \geq \overline{0} \wedge X > \overline{0}.$$

Since transitions in T_1 are live in $(N_2, M \upharpoonright ({}^\bullet T_1 \cup T_1^\bullet))$, there exists a firable transition sequence including all transitions in T_1. After firing this transition sequence, all minimal circuits of N_2 are marked. Therefore, according to Lemma 3.7 we draw a conclusion that $(N_2, M \upharpoonright ({}^\bullet T_1 \cup T_1^\bullet))$ is live, i.e., t_0 is live in $(N_2, M \upharpoonright ({}^\bullet T_1 \cup T_1^\bullet))$, which leads to a contradiction. Therefore, the sufficiency of Theorem 3.8 holds. $\square$

The liveness problem of normal Petri nets is also NP-complete [15].

Question

Try to provide a proof of this conclusion of complexity.

3.3.2 Weakly Persistent Net

3.3.2.1 Definition of Weakly Persistent Net

Chapter 1 introduces persistency. A looser property is called *weak persistency*. Formally speaking, a Petri net (N, M_0) is *weakly persistent* if for any reachable marking $M \in R(N, M_0)$, any transition t and any transition sequence σ with $\psi(\sigma)(t) = 0$, if we have $M[t\rangle$ and $M[\sigma\rangle$, then there exists a re-arrangement σ' of σ such that $M[t\sigma'\rangle$. Yamasaki [17] defines a subclass of nets called *weakly persistent nets* and proves that for such a net, the related net system is weakly persistent for any initial marking, and vice versa.

Definition 3.6 (*Weakly Persistent Net*) A net $N = (P, T, F)$ is called a *weakly persistent net* if for each circuit u there exists a trap Q such that

$$Q \subseteq P_u \wedge \Lambda^+(Q) \cap \Lambda(P_u) = \emptyset$$

where P_u is the set of places in u,

$$\Lambda^+(Q) \triangleq \{t \in T \mid t \in {}^\bullet Q \wedge t \notin Q^\bullet\},$$

and

Fig. 3.8 Two weakly persistent nets in which each circuit contains a trap $\{p_1, p_3\}$ with $\Lambda^+\{p_1, p_3\} = \emptyset$

$$\Lambda(P_u) \triangleq \{t \in T \mid t \in {}^\bullet P_u \wedge t \in P_u^\bullet\}.$$

A weakly persistent net with an initial marking is called a *weakly persistent Petri net*.[4]

Question

Try to design an algorithm to decide if a net is weakly persistent or not and analyse its complexity.

Example 3.7 None of the two nets in Fig. 3.7 is a weakly persistent net since for the circuit $u = p_1 t_1 p_2 t_2 p_3 t_3 p_4 t_4 p_1$, the transition t_3 belongs to both $\Lambda^+(Q) = \{t_3\}$ and $\Lambda(P_u) = \{t_1, t_2, t_3, t_4\}$ where $Q = \{p_1, p_4\}$ and $P_u = \{p_1, p_2, p_3, p_4\}$. Figure 3.8 shows two weakly persistent nets.

3.3.2.2 Reachability of Weakly Persistent Net

Obviously, each weakly persistent net is a normal net. Note that $\Lambda^+(Q)$ and $\Lambda(P_u)$ in Definition 3.6 can be extended to any place set X in a net:

$$\Lambda^+(X) \triangleq \{t \in T \mid t \in {}^\bullet X \wedge t \notin X^\bullet\},$$

$$\Lambda(X) \triangleq \{t \in T \mid t \in {}^\bullet X \wedge t \in X^\bullet\}.$$

---

[4] The concept of *weakly persistent Petri net* in this definition is not equivalent to that in the above paragraph. When a net is not a weakly persistent net, it with some initial marking possibly satisfies the property of weak persistency, but it with some other initial marking does not. For example in Fig. 3.7b, the Petri net satisfies the weakly persistent property, but the net is not weakly persistent. If the initial marking is $M_0 = \{p_2, p_3, p_5\}$, then we have that $M_0[t_2\rangle$ and $M_0[t_3t_4t_1t_2\rangle$ but $\neg M_0[t_2\sigma$ for each re-arrangement σ of $t_3t_4t_1$, i.e., this Petri net is not weakly persistent when it is equipped with the initial marking $\{p_2, p_3, p_5\}$. Hence, when we talk about a weakly persistent Petri net, we assume that the underlying net must be weakly persistent.

Given a weakly persistent net $N = (P, T, F)$, a marking M and a nonnegative integer vector X indexed by T, the subnet $(N, M) \upharpoonright X \triangleq (T', P', F', M')$ is a *projection* of (N, M) over X where

- $T' = \{t \in T \mid X(t) > 0\}$,
- $P' = {}^\bullet T' \cup T'^\bullet$,
- $F' = F \cap ((P' \times T') \cup (T' \times P'))$, and
- $M' = M \upharpoonright P'$.

Theorem 3.9 ([18]) *Let $(N, M_0) = (P, T, F, M_0)$ be a weakly persistent Petri net. A marking M is reachable in (N, M_0) if and only if there is a nonnegative integer vector X satisfying*

1. *$M = M_0 + X \cdot [N]$, and*
2. *each siphon circuit of $N \upharpoonright X$ is marked at M_0.*

Proof (Necessity) This decision condition as a necessary condition also holds for an arbitrary Petri net. When a marking is reachable in a Petri net, then the firable transition sequence σ leading to the marking corresponds to a firing count vector $\psi(\sigma)$ that is such a nonnegative integer vector. According to the introduction in Sect. 2.1 in Chap. 2, we know that $\psi(\sigma)$ satisfies Condition 1. It also satisfies Condition 2. (By contradiction) If a siphon in the subnet w.r.t. $\psi(\sigma)$ is unmarked at the initial marking, then the transitions that occur in σ and are related to this siphon cannot be enabled forever in the subnet system, which obviously contradicts the fact that they can be fired in the run of σ.

(Sufficiency) To prove the sufficiency, we only need to prove the following three conclusions [18]:

1. Given a weakly persistent net N and a nonnegative integer vector X indexed by transitions of N, then $N \upharpoonright X$ is still a weakly persistent net.
2. Given a weakly persistent Petri net (N, M) in which each siphon circuit is marked at M, then there is a transition t such that $M[t\rangle$.
3. Given a firable transition sequence σ in (N, M_0) satisfying

$$X' = X - \psi(\sigma) \geq \overline{0},$$

then each siphon circuit in $(N, M') \upharpoonright X'$ is still marked at M' where $M_0[\sigma\rangle M'$.

The first conclusion is obvious.

The second conclusion holds for any Petri net (N, M) with

$$M + X \cdot [N] \geq \overline{0} \wedge X > \overline{0}.$$

If every transition is disabled for such a Petri net but every siphon circuit is marked, then a subset of those unmarked places form a siphon circuit which contradicts the fact that every siphon circuit is marked.

We assume that the third conclusion does not hold, i.e., there is a siphon circuit u which is unmarked in $(N, M') \upharpoonright X'$. Let $Q \subseteq P_u$ be such a trap satisfying the condition of Definition 3.6. Since P_u is a siphon unmarked in $(N, M') \upharpoonright X'$, we have that $\Lambda^+(Q) = \emptyset$ and Q is unmarked in $(N, M') \upharpoonright X'$. There are two cases:

- $\Lambda^+(Q) = \emptyset$ in $(N, M_0) \upharpoonright X$. If this case holds, then Q is both a siphon and a trap in $(N, M_0) \upharpoonright X$ and thus is unmarked in $(N, M_0) \upharpoonright X$. This contradicts the fact that each siphon circuit is marked in $(N, M_0) \upharpoonright X$.
- $\Lambda^+(Q) \neq \emptyset$ in $(N, M_0) \upharpoonright X$. If this case holds, then firing transition sequence σ puts at least one token into Q. A contradiction is also yielded.

The above two cases cannot occur, which means that the assumption does not hold. Hence, the third conclusion holds.   □

Example 3.8 If we equip the weakly persistent net in Fig. 3.8a with the initial marking $M_0 = \bar{0}$ (i.e. every place has no token), then the marking $M = \{p_2\} = [0, 1, 0]$ is not reachable obviously. Although there is a vector $X = [2, 1, 2, 0]$ indexed by t_1, t_2, t_3, and t_4 such that

$$[0, 1, 0] = [0, 0, 0] + [2, 1, 3, 0] \cdot \begin{bmatrix} -1 & 1 & 1 \\ -1 & -1 & 1 \\ 1 & 0 & -1 \\ 0 & -1 & 0 \end{bmatrix},$$

each siphon circuit of $(N, M_0) \upharpoonright X$ is not marked at M_0, as shown in Fig. 3.8b.

The above example shows that the second condition in Theorem 3.9 is necessary to decide reachability. However, the conclusion in Theorem 3.9 does not necessarily hold for a normal Petri net even when the two conditions both hold.

Example 3.9 In Fig. 3.7b, the marking $M = [1, 1, 0, 0, 1, 0, 0] = \{p_1, p_2, p_5\}$ is not reachable from the initial marking $M_0 = [0, 1, 0, 0, 1, 0, 0] = \{p_2, p_5\}$ in the normal Petri net. However, there is $X = [3, 1, 1, 4, 1, 1]$ such that $M = M_0 + X \cdot [N]$, and each siphon circuit in $(N, M_0) \upharpoonright X$ is marked at M_0 where $(N, M_0) \upharpoonright X$ is (N, M_0) itself.

> The reachability problem of weakly persistent Petri nets is still NP-complete.

To the best of our knowledge, we have not seen the complexity work of the reachability problem of weakly persistent Petri nets, but it is not difficult to draw this conclusion. First, we know that each net without any circuit is a weakly persistent net according to Definition 3.6. Additionally, referring to the polynomial-time reduction from the satisfiability problem of 3-SAT to the soundness problem of workflow nets without circuit [19–21], we easily reduce the satisfiability problem of 3-SAT to the reachability problem of weakly persistent Petri nets (without circuit), which means that the reachability problem of weakly persistent Petri nets is NP-hard. Furthermore, the NP algorithm of deciding the reachability of normal Petri nets in [15] is still suitable for the reachability problem of weakly persistent Petri nets. Hence, this problem is NP-complete.

3.3.2.3   Liveness of Weakly Persistent Net

Theorem 3.10 ([5]) *A weakly persistent Petri net is live if and only if every siphon contains an initially marked trap.*

If each siphon in a weakly persistent Petri net contains an initially marked trap, then each siphon is marked at each reachable marking. Additionally, each weakly persistent Petri net is a normal Petri net. Hence, according to Theorem 3.8 we know that the sufficiency of Theorem 3.10 holds. The necessity can be derived from the following conclusion:

Lemma 3.9 ([5]) *If S is a minimal siphon in a weakly persistent net such that*

$$^\bullet S \neq \emptyset,$$

then S is also a trap.

Proof Since $^\bullet S \neq \emptyset$, we have that there is a circuit u such that $P_u = S.$[5] Therefore, there must exist a trap $Q \subseteq S$ such that

$$\Lambda^+(Q) \cap \Lambda(S) = \emptyset$$

according to the definition of weakly persistent net. If we can prove that for each non-empty proper subset Q' of S, the following always holds:

$$Q'^\bullet \subseteq \, ^\bullet Q' \Rightarrow \Lambda^+(Q') \cap \Lambda(S) \neq \emptyset,$$

then we can derive that $Q = S$ is exactly the trap we want. First, given such a trap $Q' \subset S$, we observe

$$Q'^\bullet \neq \, ^\bullet Q'.$$

(By contradiction) If $Q'^\bullet = \, ^\bullet Q'$, we know that Q' is also a siphon, which contradicts the fact that S is a minimal siphon. Therefore, we have

$$Q'^\bullet \subset \, ^\bullet Q',$$

i.e.,

---

[5] It can be derived from such a conclusion: given a minimal siphon S in a net $N = (P, T, F)$, the subnet $(S, \, ^\bullet S, F \cap ((S \times \, ^\bullet S) \cup (^\bullet S \times S)))$ is strongly connected [1, 22].

$$\Lambda^+(Q) \neq \emptyset.$$

In addition, for each $t \in \Lambda^+(Q)$, it also belongs to $^\bullet S$ since $Q' \subset S$. In consequence, it also belongs to $S^\bullet$ since S is a siphon. Therefore, for each $t \in \Lambda^+(Q)$, we have

$$t \in \Lambda(S),$$

i.e.,

$$\Lambda^+(Q') \cap \Lambda(S) \neq \emptyset.$$

Therefore, a minimal siphon itself is also a trap in a weakly persistent net when its input set is not empty. ☐

Now that each minimal siphon is a trap in a weakly persistent net and every siphon contains a minimal one, we can simplify Theorem 3.10 as follows:

Corollary 1 *A weakly persistent Petri net is live if and only if every minimal siphon is initially marked.*

The above conclusion ensures a polynomial-time algorithm to decide the liveness of weakly persistent Petri net [23]. We can delete all initially marked places (as well as all associated arcs) from a given weakly persistent Petri net, and check if there is a siphon in the remained net. This check can be started from those transitions without input. Obviously, it can be done in $\mathcal{O}(|P| + |T| + |F|)$.

3.4 Application

Figure 3.9 shows a bill of materials of a manufacturing system in which 2 types of products U_1 and U_2 are produced from 3 types of raw parts W_1, W_2 and W_3 through 5 different types of machines V_j ($j \in \mathbb{N}_5^+$) and 2 assemble stations A_1 and A_2. U_1 is the assemble of W_1 and W_2 and can use 2 different assemble lines: (1) after W_1

Fig. 3.9 A bill of materials of a flexible manufacturing systems [24]

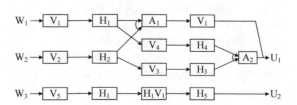

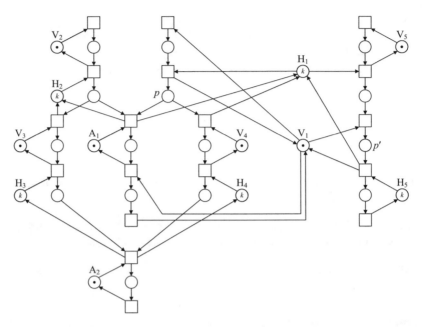

Fig. 3.10 The Petri net modeling the manufacturing system in Fig. 3.9

and W_2 are respectively processed by V_1 and V_2, they are assembled in A_1 and then will be processed again by V_1; or (2) W_1 is processed by V_1 and V_4 in turn, W_2 is processed by V_1 and V_3 in turn, and then they are assembled in A_2. U_2 is made of W_3 first processed by V_5 and then by V_1. A part processed by a machine will be put into a buffer only if the buffer is not full. Here we assume that each machine can process one part at a time, each assemble station can assemble two parts from two different buffers and the capacity of each buffer is k ($k >> 1$). We also assume that there are enough raw parts and the final products can be moved in time, i.e.. it is reasonable that a model does not take into account the buffers storing the raw parts and the final products. Therefore, we can use the Petri net in Fig. 3.10 to model this manufacturing system.

Here, machine V_1 and buffer H_1 are shared by multiple processes and thus possibly results in a circular wait situation (i.e., (partial) deadlock). For instance, the situation, where H_1 stores k parts W_3 that have processed by machine V_5 and machine V_1 is used to process a part W_1, is a partial deadlock at which the siphon $\{p, p', H_1, V_1\}$ becomes empty. Here we can use the deadlock-prevention policy proposed in [25] to split the buffer H_1 into two sub-buffers: one is used to store W_1 and another one stores W_3 so that the circular wait situation is destroyed. Additionally, since each machine uses a fixed buffer when it processes a part, we can combine some transitions, as shown in Fig. 3.11, such that the behaviours of the system are preserved while being convenient to analyse it. As shown in Fig. 3.11, the net is a weakly persistent net

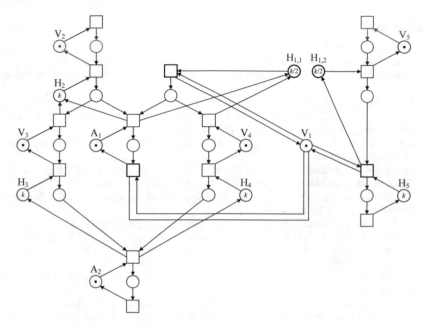

Fig. 3.11 Buffer H_1 is split into two ones $H_{1,1}$ and $H_{1,2}$, and this net is a weakly persistent net

since we can easily check that each circuit is a trap circuit. Therefore, the system obtained by splitting the buffer H_1 is live and thus no (partial) deadlock can take place.

3.5 Summary and Further Reading

This chapter introduces some classical subclasses of Petri nets. For more knowledge of state machine, marked graph, free-choice net and asymmetric-choice net, one can further read the book [1]. As for normal net and weakly persistent net, one can find more results in [2–5, 14, 15, 17, 18, 23]. Just as they are of some beautiful structures, some sufficient and necessary conditions can be provided to decide some properties such as liveness and reachability based on these structures. State equation, siphon and trap play an important role in these conditions. There are some other structure-based analysis methods for some subclasses of Petri nets, e.g. rank theorem, T-/P-invariant and circular wait chain [1, 26–31].

Among these subclasses of Petri nets, asymmetric-choice nets are worth being studied further [13, 32–34]. On the one hand, they can model more complex process structures such as interaction and loop in applications. On the other hand, structure-based analysis methods have not be discovered completely in theory so far.

References

1. Desel, J., Esparza, J.: *Free Choice Petri Nets*. Cambridge University Press (1995)
2. Murata, T.: Petri nets: properties, analysis and application. *Proceedings of IEEE* 77: 541–580 (1989)
3. Yamasaki, H.: Normal Petri nets. *Theoretical Computer Sciences* 31: 307–315 (1984)
4. Liu, G.J., Jiang, C.J., Chao, D.Y.: . A necessary and sufficient condition for the liveness of normal nets. *The Computer journal* 54: 157–163 (2011)
5. Liu, G.J., Jiang, C.J.: On conditions for the liveness of weakly persistent nets. *Information Processing Letters* 109: 967–970 (2009)
6. Georgiadis, L., Italiano, G.F., Parotsidis, N.: Strong connectivity in directed graphs under failures with applications. *SIAM Journal on Computing* 49: 865–926 (2020)
7. Simon, K.: An improved algorithm for transitive closure on acyclic digraphs. *Theoretical Computer Sciences* 58: 325–346 (1988)
8. Lu, W., Zhao, Q., Zhou, C.: A parallel algorithm for finding all elementary circuits of a directed graph. In: *Proceedings of the 37th Chinese Control Conference* pp. 3156–3161 (2018)
9. Johnson DB. Finding all the elementary circuits of a directed graph. *SIAM Journal on Computing* 4: 77–84 (1975)
10. Best, E.: Structural theory of Petri nets: the free choice hiatus. *Lecture Notes in Computer Science* 254: 168–206 (1987)
11. Wu, Z.H.: *Introduction to Petri Nets*. China Machine Press, Beijing (2006) (in Chinese)
12. Jones, N.D., Landweber, L.H., Lien, Y.E.: Complexity of some problems in Petri nets. *Theoretical Computer Science* 4: 277–299, (1977)
13. Liu, G.J.: PSPACE-completeness of the soundness problem of safe asymmetric-choice workflow nets. In: *the 41st International Conference on Application and Theory of Petri Nets and Concurrency*, pp. 196–216 (2020)
14. Howell, R.R., Rosier, L.E., Yen, H.C.: Normal and sinkless petri nets. *Journal of Computer and System Sciences* 46: 1–26 (1993)
15. Ohta, A., Tsuji, K.: Computational complexity of liveness problem of normal petri net. *IEICE Transactions on Fundamentals of Electronics, Communications and Computer Sciences* E92-A, 2717–2722 (2009)
16. Barkaoui, K., Couvreur, J.M., Klai, K.: On the Equivalence Between Liveness and Deadlock-Freeness in Petri Nets. In: *the 26th International Conference on Applications and Theory of Petri Nets*, pp. 90–107 (2005)
17. Yamasaki, H.: On weak persistency of Petri nets. *Information Processing Letters* 13: 94–97, (1981)
18. Hiraishi, K., Ichikawa, A.: On structural conditions for weak persistency and semilinearity of Petri nets. *Theoretical Computer Science* 93: 185–199 (1992)
19. Liu, G.J., Liu, Y., Sun, J., Dong, J.S.: Complexity of the soundness problem of bounded workflow nets. In: *the 33rd international conference on Application and Theory of Petri Nets*, pp. 92–107 (2012)
20. Liu, G.J., Liu, Y., Sun, J., Dong, J.S.: Complexity of the soundness problem of workflow nets. *Fundamenta Informaticae* 131: 81–101, (2014)
21. Liu, G.J.: *Primary Unfolding of Petri Nets: A Model Checking Approach of Concurrent Systems*. Beijing: China Science Press (2020) (in Chinese)
22. Tanimoto, S., Yamauchi, M., Watanabe, T.: Finding minimal siphons in general petri nets. *IEICE Transactions on Fundamentals of Electronics, Communications and Computer Sciences* E79-A: 1817–1824 (1996)
23. Ohta, A., Liu, G.J., Tsuji, K., Jiang, C.J., Chao, D.: Polynomial time solvability of liveness problem of structurally weakly persistent net, *IEICE Technical Report* 110: 127–132 (2010)
24. Jeng, M.D., Xie, X.L., Peng, M.Y.: Process nets with resources for manufacturing modeling and their analysis. *IEEE Transactions on Robotics and Automation* 18: 875–889 (2003)

25. Liu, G.J., Jiang, C.J., Chen, L.J., Wu, Z.H.: Two types of extended RSNBs and their application in modelling flexible manufacturing systems. *International Journal of Advanced Manufacturing Technology* 45: 573–582 (2009)

26. Liu, G.J., Jiang, C.J., Zhou, M.C., Ohta, A.: The liveness of WS^3PR: complexity and decision. *IEICE Transactions on Fundamentals of Electronics, Communications and Computer Sciences* E96-A: 1783–1793 (2013)

27. Liu, G.J., Chen, L.J.: Deciding the liveness for a subclass of weighted Petri nets based on structurally circular wait. *International Journal of Systems and Sciences* 47: 1533–1542 (2016)

28. Silva, M., Teruel, E., Colom, J.: Linear algebraic and linear programming techniques for the analysis of place/transition net systems. *Lecture Notes on Computer Sciences* 1491: 309–373 (1996)

29. Xing, K.Y., Wang, F., Zhou, M.C., Lei, H., Luo, J.: Deadlock characterization and control of flexible assembly systems with Petri nets. *Automatica* 87: 358-364 (2018)

30. Wimmel, H.: Presynthesis of bounded choice-free or fork-attribution nets. *Information and Computation* 271: 104482-1–104482-20 (2020)

31. Best, E., Hujsa, T., Wimmel, H.: Sufficient conditions for the marked graph realisability of labelled transition systems. *Theoretical Computer Science* 750: 101–116 (2018)

32. Best, E., Devillers, R., Erofeev, E., Wimmel, H.: Target-oriented Petri net synthesis. *Fundamenta Informaticae* 175: 97–122 (2020)

33. Matsubara, S., Yamaguchi, S., Ahmadon, M.A.B.: Generating and analyzing data set of workflow-Nets. In: *the 8th International Symposium on Computing and Networking Workshops*, pp. 471–473 (2020).

34. Tredup, R.: Synthesis of structurally restricted *b*-bounded Petri nets: complexity results. In: *the 13th International Conference on Reachability Problems*, pp. 202–217 (2019)

Chapter 4
Petri Nets Modeling Message Passing and Resource Sharing

4.1 Workflow Nets

Workflow nets are defined by Dr. Wil van der Aaslt [1] and can well model the execution logics of workflows.

4.1.1 Soundness and Weak Soundness of Workflow Nets

A workflow net has no too much constrains on its structures except two special places (i.e., one without input and one without output) as well as connectivity (i.e., each node occurs in a directed path between the two special places).

Definition 4.1 (*Workflow Net*) A net $N = (P, T, F)$ is called a *workflow net* if the following conditions hold:

1. N has two special places i and o where $i \in P$ satisfying ${}^\bullet i = \emptyset$ is its *source place* and $o \in P$ satisfying $o^\bullet = \emptyset$ is its *sink place*, and
2. the *trivial extension* $\overline{N} \triangleq (P, T \cup \{t_0\}, F \cup \{(t_0, i), (o, t_0)\})$ of N is strongly connected where $t_0 \notin T$.

For a correct workflow, its behaviours should fulfil that any run should terminate eventually and correctly. "eventual termination" means that *the sink place o will always be marked in any run*, while "correct termination" means that *when o is marked, all the other places are unmarked*. Such a run is usually called "properly terminate". To this end, Aalst [1] proposes the concept of *soundness* as follows:

Definition 4.2 (*Soundness of Workflow Net*) A workflow net $N = (P, T, F)$ is *sound* if the following conditions hold:

© The Author(s), under exclusive license to Springer Nature Singapore Pte Ltd. 2022
G. Liu, *Petri Nets*,
https://doi.org/10.1007/978-981-19-6309-4_4

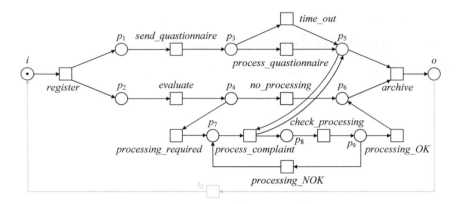

Fig. 4.1 A workflow net modelling the processing of complaints in some company [1]. If the transition t_0 and the two arcs associated with it are considered, then the resulting net is its trivial extension

1. for each $M \in R(N, \{i\})$ we have $\{o\} \in R(N, M)$,
2. for each $M \in R(N, \{i\})$ if $M \geq \{o\}$ then $M = \{o\}$, and
3. for each $t \in T$ there exists $M \in R(N, \{i\})$ such that $M[t\rangle$.

In fact, the second condition in Definition 4.2 is implied by the first one due to the arbitrary-ness of M in the first condition [1]. The third one means that each action has a potential chance to happen.

Question

Try to provide a rigour proof that the second condition is implied by the first one.

Example 4.1 Figure 4.1 shows a workflow net modelling the processing of con-sumers' complaints to the products of a company. Using its reachability graph, we easily check that it is sound.

Generally, $\{i\}$ and $\{o\}$ are called the *initial* and *target* markings of a workflow net, respectively. Additionally, a safe (respectively bounded, unbounded) workflow net means that the net with the initial marking is safe (respectively bounded, unbounded).

If the third requirement is removed from Definition 4.2, i.e., some transitions may have no enabled chance, then this definition becomes *weak soundness*.

Definition 4.3 (*Weak Soundness of Workflow Net*) A workflow net N is *weakly sound* if the following condition holds:

$$\forall M \in R(N, \{i\}) : \{o\} \in R(N, M).$$

The definition of (weak) soundness is based on reachability (and thus can be
checked through reachability graphs), but actually they are closely related to
liveness, boundedness, deadlock, and livelock.

Theorem 4.1 ([1]) *A workflow net N is sound if and only if* $(\overline{N}, \{i\})$ *is both live and
bounded where* $\overline{N}$ *is the trivial extension of N.*

Proof The necessity is obvious. (Sufficiency) We denote t_0 as the transition that
connects i and o in $\overline{N}$. Since $(\overline{N}, \{i\})$ is live, we have that for each marking $M \in
R(N, \{i\})$, there must exist a marking $M' \in R(N, M)$ such that

$$M'(o) > 0,$$

because the transition t_0 is live in $(\overline{N}, \{i\})$. Now we show that M' does not mark any
other place rather than o. (By contradiction) We assume that there is a place $p \neq o$
that is also marked at M', and let σ be the transition sequence in N such that

$$\{i\}[\sigma\rangle M'.$$

Then, the transition sequence σt_0 can be repeatedly fired in $(\overline{N}, \{i\})$ and produce
more and more tokens into p. This means that $(\overline{N}, \{i\})$ is unbounded, which contradicts its boundedness. Therefore, when o is marked, other places cannot be marked
simultaneously for a sound workflow net.   □

Here, we re-define deadlock for workflow nets: a reachable marking $M \in
R(N, \{i\})$ is called a *deadlock* of $N = (P, T, F)$ if the following holds:

$$M \neq \{o\} \wedge \forall t \in T : \neg M[t\rangle.$$

Similarly, we define livelock as follows: a reachable marking $M \in R(N, \{i\})$ is
called a *livelock* of N if the following holds:

$$\{o\} \notin R(N, M) \wedge \forall M' \in R(N, M), \exists t \in T : M'[t\rangle.$$

Obviously, they are defined around the target marking. A deadlock means such a
situation where the target marking has not been reached but any transition is disabled.
A livelock means such a situation where the target marking will not be reachable but
there is always a firable transition. Obviously, any successor of a livelock is still a
livelock.

The following conclusion is obviously derived from their definitions:

Theorem 4.2 ([2]) *A workflow net N is weakly sound if and only if* $(N, \{i\})$ *has
neither deadlock nor livelock.*

Although the soundness problem of workflow nets can be solved according to the liveness and boundedness of their trivial extensions, the complexity of this problem is very high. For example, for safe workflow nets, the complexity of their soundness problem is PSPACE-complete [3]^a; and even for safe acyclic workflow nets, it is still co-NP-complete [4]. Fortunately, for free-choice work-flow nets, their soundness can be decided in polynomial time [1]. However, for asymmetric-choice workflow nets, this problem is not easy; even for safe asymmetric-choice workflow nets, their soundness problem is still PSPACE-complete [5].

---

^{a} A safe workflow net means that this workflow net with the initial marking $\{i\}$ is safe. Additionally, an acyclic workflow net means that this workflow net has no circuit. It is similar to asymmetric-choice workflow nets, free-choice workflow nets, and so on.

Question

Try to prove these conclusions of complexity.

In what follows, we discover that for free-choice workflow nets, their weak soundness problem is equivalent to their soundness problem.

4.1.2 Equivalence Between Soundness and Weak Soundness in Free-choice Workflow Nets

Theorem 4.3 ([6]) *A free-choice workflow net is weakly sound if and only if it is sound.*

Proof The sufficiency can be derived directly according to their definitions. The proof of necessity uses the contradiction method. We assume that there exists a free-choice workflow net $N = (P, T, F)$ that is weakly sound but not sound. Then, there is a transition $t \in T$ such that

$$\forall M \in R(N, \{i\}) : \neg M[t\rangle,$$

i.e., t has never an enabled chance. There are two cases about the input(s) of t:

- $|{}^\bullet t| = 1$. For this case, this unique input place of t is unmarked forever in $(N, \{i\})$.
- $|{}^\bullet t| > 1$. For this case, we know that for each $p \in {}^\bullet t$:

$$p^\bullet = \{t\},$$

since N is a free-choice net. Now we can claim that every input place of t is unmarked forever in $(N, \{i\})$. If some one (say p) is marked, then the token in it cannot be removed since its output is t but t has never an enabled chance. In consequence, when o is marked, p is also marked, which contradicts the definition of weak soundness.

In a word, every input place of t cannot be marked in the run of $(N, \{i\})$ if t has never an enabled chance. Hence, we know that any transition belonging to $^{\bullet\bullet}t$ has no enabled chance in the run of $(N, \{i\})$ either, and thus their input places cannot be marked forever. Consequently, we can derive that i is unmarked forever, which contradicts the fact that it is mark at the initial marking. Therefore, when a free-choice workflow net is weakly sound, it is sound.   □

4.1.3 Equivalence Between Soundness and Weak Soundness in Acyclic Asymmetric-Choice Workflow Nets

The equivalence conclusion in the above section also holds for the acyclic asymmetric-choice workflow nets.

Theorem 4.4 ([7]) *An acyclic asymmetric-choice workflow net is weakly sound if and only if it is sound.*

Proof We also use the contradiction method to prove the necessity. We assume that there is an asymmetric-choice workflow net $N = (P, T, F)$ that is weakly sound but not sound, and thus there is a transition $t \in T$ such that

$$\forall M \in R(N, \{i\}) : \neg M[t\rangle.$$

We denote $p_1, p_2, \ldots, p_k$ as all input places of t where $k \geq 1$. Just stated in Sect. 3.2 in Chap. 3, there is an inclusion relation among $p_1^\bullet, p_2^\bullet, \ldots,$ and $p_k^\bullet$ [8]. Without loss of generality, we let

$$p_1^\bullet \subseteq p_2^\bullet \subseteq \cdots \subseteq p_k^\bullet.$$

In what follows, we first draw the following conclusion from the above assumption:

$$\forall M \in R(N, \{i\}) : M(p_1) = 0.$$

(By contradiction) If there exists a marking $M \in R(N, \{i\})$ such that

$$M(p_1) > 0,$$

Fig. 4.2 An
asymmetric-choice workflow
net with a circuit that is
weakly sound but unsound.
The transitions in the circuit
have no enabled chance in
the run

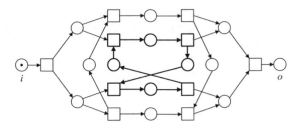

then there must exist a transaction (say t_1) that can remove the tokens from p_1 since
$(N, \{i\})$ is weakly sound. Due to the asymmetric-choice property and $p_1^\bullet \subseteq p_2^\bullet \subseteq$
$\cdots \subseteq p_k^\bullet$, we know that

$$\{p_1, p_2, \ldots, p_k\} \subseteq {}^\bullet t_1.$$

This means that when t_1 is enabled, t is also enabled. That is, t has an enabled chance
in $(N, \{i\})$, which contradicts the assumption that t is disabled forever.

Now that p_1 is unmarked forever in the run of $(N, \{i\})$, every input transition of
p_1 has no enabled chance in the run of $(N, \{i\})$. Because N is acyclic, we repeat the
above process and can finally find a transition t' and a place p' such that

$$\{i, p'\} \subseteq {}^\bullet t' \wedge \forall M \in R(N, \{i\}) : M(p') = 0.$$

Starting from p', we can repeat the above process. Since N is acyclic, we can find
another transition $t'' \neq t'$ and another place $p'' \neq p'$ such that

$$\{i, p''\} \subseteq {}^\bullet t'' \wedge \forall M \in R(N, \{i\}) : M(p'') = 0.$$

Repeating the above process, we can find infinitely many pairs of such transitions
and places, which contradicts the finiteness of the number of transitions and places
of N.   □

> Theorem 4.4 does not necessarily hold for asymmetric-choice workflow nets
> with circuits.

For instance, the asymmetric-choice workflow net in Fig. 4.2 is weakly sound but
unsound, because these transitions in the unique circuit have no enabled chance.

4.1.4   k-Soundness

Hee et al. [9] extend the concept of soundness to *k-soundness* representing the correctness of the (simultaneous) run of k cases in a workflow. k tokens are put into the

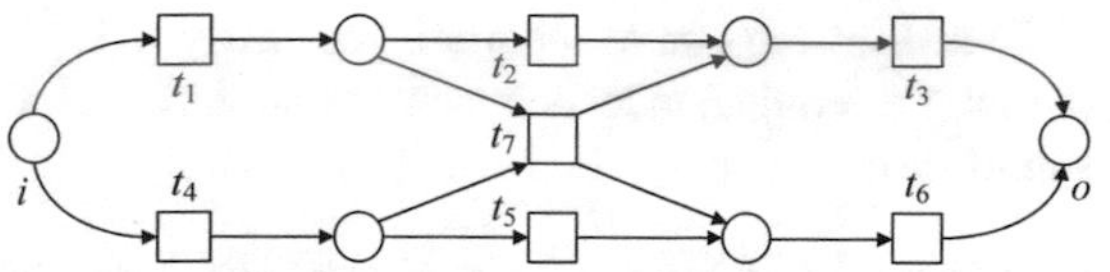

Fig. 4.3 A workflow net
that is 2-sound but not sound

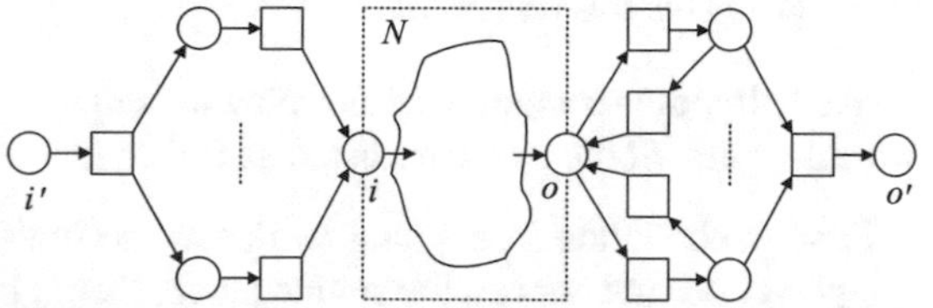

Fig. 4.4 Deciding
k-soundness can be
transferred into deciding
soundness

source place as the initial marking and k-soundness requires that the k tokens are
always moved into the sink place properly.

Definition 4.4 (*k-Soundness of Workflow Net*) A workflow net $N = (P, T, F)$ is
k-sound for a given $k \in \mathbb{N}^+$ if the following conditions hold:

1. for each $M \in R(N, \{k \cdot i\})$ we have $\{k \cdot o\} \in R(N, M)$, and
2. for each $t \in T$ there exists $M \in R(N, \{k \cdot i\})$ such that $M[t\rangle$.[1]

Obviously, soundness in Definition 4.2 is 1-soundness defined here.

> When a workflow net is not sound, it is possibly k-sound for some k.

For instance, the workflow net in Fig. 4.3 is not sound since t_7 has any enabled
chance in $(N, \{i\})$, but it is 2-sound.

> In theory, however, k-soundness does not ensure more or stronger design
> requirements than soundness.

How to understand the above assertion? Given a workflow net, when we check
if it is k-soundness for a given k, the checking can be transferred into checking the
soundness of another workflow net that is just a simple extension of the original one,
as shown in Fig. 4.4. Here, we use $k + 1$ transitions to produce k tokens into the
original source place, and $2k + 1$ transitions to delete k tokens from the original sink
place. Therefore, we have the following conclusion:

Theorem 4.5 ([6]) *A workflow net is k-sound for a given positive integer k if and
only if the constructed workflow net as shown in Fig. 4.4 is sound.*

---

[1] In [9], k-soundness does not require that every transition has an enabled chance; in other words,
it corresponds to weak k-soundness. For consistency, we add this requirement in this book.

Based on the above theorem, we easily extend the conclusion in Theorem 4.1 (about the equivalence between soundness and liveness-and-boundedness) to k-soundness:

Corollary 4.1 *A workflow net N is k-sound for a given positive integer k if and only if $(\overline{N}, \{k \cdot i\})$ is live and bounded.*

Actually, the structure of a net play an important role in the design of a sound workflow net. Some cases are listed as follows:

- First, each siphon in a sound workflow net must contain the source place. If a siphon does not contain the source place, then it keeps unmarked at any reachable marking and thus the transitions associated with it have no enabled chance, e.g. the asymmetric-choice workflow net in Fig. 4.2 is not k-sound for any positive integer k since it contains a siphon without the source place.
- Second, any trap in a sound workflow net must contains the sink place. If there is a trap that does not contain the sink place, then tokens can be left in the trap when the source place is marked.
- Additionally, a sound workflow net means that every complete run (i.e., the token reaches the sink place from the source one) corresponds to a T-invariant in $\overline{N}$ (i.e., the transitions in the run plus the transition t_0 connecting i and o). Therefore, a sound workflow net should be covered by T-invariant.

According to Theorems 3.6 and 4.1, we easily draw the following conclusion:

Corollary 4.2 *An asymmetric-choice workflow net is k-sound for a given positive integer k if and only if $(\overline{N}, \{k \cdot i\})$ is bounded and each siphon is marked at each reachable marking.*

4.2   Inter-organisational Workflow Nets

A workflow net generally describes the business process in an enterprise. With the expansion of businesses of different enterprises, more and more businesses cross organisations; and with the development of technology such as web service, cross-organisational enterprise information systems can be developed. Inter-organisational workflow nets can be used to model and analyse business processes of such systems.

4.2.1   Compatibility

In [10], an inter-organisational workflow net considers both synchronous and asynchronous communications among workflow nets. Here we only consider the asynchronous pattern.

Definition 4.5 (*Inter-organisational Workflow Net*)   A net $N^{\otimes} \triangleq (N_1, \ldots, N_m,$ $P, F)$ is called an *inter-organisational workflow net* if the following conditions hold:

1. $N_1 = (P_1, T_1, F_1), \ldots,$ and $N_m = (P_m, T_m, F_m)$ are pairwise disjoint workflow nets, called *basic workflow nets* for convenience, where $m \geq 1$,
2. P is a finite set of *channel places* such that $P \cap P_j = \emptyset$ for each $j \in \mathbb{N}_m^+$,
3. $F \subseteq \bigcup_{j=1}^{m}(P \times T_j) \cup (T_j \times P)$ is a set of arcs connecting different workflow nets via channel places, and
4. for each $p \in P$ there exist two different positive integers $j \in \mathbb{N}_m^+$ and $k \in \mathbb{N}_m^+$ such that ${}^{\bullet}p \subseteq T_j, p^{\bullet} \subseteq T_k, {}^{\bullet}p \neq \emptyset$, and $p^{\bullet} \neq \emptyset$.

Each channel place is used only by two fixed basic workflow nets according to the fourth condition in Definition 4.5. In other words, two different basic workflow nets cannot send messages into the same channel place; similarly, they cannot take messages from the same channel place either; certainly, they may use multiple channel places to communicate.

Definition 4.6 (*Compatibility of Inter-organisational Workflow Net*)   Let $N^{\otimes} =$ $(N_1, \ldots, N_m, P, F)$ be an inter-organisational workflow net, and i_j and o_j be the source and sink places of N_j, respectively, for each $j \in \mathbb{N}_m^+$. $N^{\otimes}$ is *compatible* if the following conditions hold:

1. for each $M \in R(N^{\otimes}, \{i_1, \ldots, i_m\})$ we have $\{o_1, \ldots, o_m\} \in R(N^{\otimes}, M)$, and
2. for each $t \in \bigcup_{j=1}^{m} T_j$ there exists $M \in R(N^{\otimes}, \{i_1, \ldots, i_m\})$ such that $M[t\rangle$.

Example 4.2 Figure 4.5 shows two inter-organisational workflow nets: (a) is compatible but (b) is not.

Inter-organisational workflow nets are equivalent to workflow nets when we consider the decision of compatibility.

If two new places i and o and two new transitions t_i and t_o are added to an inter-organisational workflow net such that

$$ {}^{\bullet}t_i = \{i\}, t_i^{\bullet} = \{i_1, \ldots, i_m\}, {}^{\bullet}t_o = \{o_1, \ldots, o_m\}, \text{ and } t_o^{\bullet} = \{o\}, $$

then the new net is a workflow net, and the inter-organisational workflow net is compatible if and only if this new workflow net is sound. Therefore, an inter-organisational workflow net may be thought of as a special workflow net. In the definition of inter-organisational workflow net, we may require that each basic workflow is sound; but due to the equivalence shown here, we consider this definition in a general form.

Similar to weak soundness, *weak compatibility* can be defined and omitted here.

Fig. 4.5 Two inter-organisational workflow nets [11]: **a** compatible but **b** incompatible

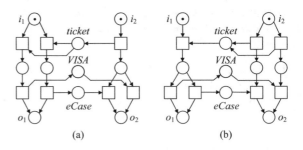

(a)   (b)

When multiple workflows are combined, obeying some interaction patterns can usually reduce the occurrence of errors [12]. And net structures and unfolding are sometimes helpful for the analysis of soundness and compatibility in order to avoid the state explosion problem [7, 13–15, 17, 18].

> Compatibility requires that each basic workflow must take part in every run, but this requirement is sometimes unnecessary. Hence, collaborative-ness is defined that permits *a part* but *not all* of basic workflows to participate in a run.

4.2.2 Collaborative-ness

Collaborative-ness means that a basic workflow in an inter-organisational workflow must terminate properly once it participate in some run and no message is left in channels. It does not require that every basic workflow must take part in every run.

Definition 4.7 (*Collaborative-ness of Inter-organisational Workflow Net*) Let $N^\otimes = (N_1, \ldots, N_m, P, F)$ be an inter-organisational workflow net, and i_j and o_j be the source and sink places of workflow net $N_j = (P_j, T_j, F_j)$, respectively, for each $j \in \mathbb{N}_m^+$. $N^\otimes$ is *collaborative* if the following conditions hold:

1. for each $M \in R(N^\otimes, \{i_1, \ldots, i_m\})$ there exists $M' \in R(N^\otimes, M)$ such that

$$M'\left(\bigcup_{j=1}^{m}\{i_j, o_j\}\right) = M'\left(\bigcup_{j=1}^{m}(P \cup P_j)\right) = m,$$

2. for each $t \in \bigcup_{j=1}^{m} T_j$ there exists $M \in R(N^\otimes, \{i_1, \ldots, i_m\})$ such that $M[t\rangle$.

   Marking M' satisfying

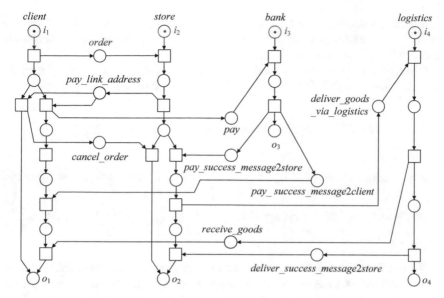

Fig. 4.6 An inter-organisational workflow net permitting a part of workflows to participate in a run. Although it is not compatible, it is collaborative

$$M'(\bigcup_{j=1}^{m}\{i_j, o_j\}) = M'(\bigcup_{j=1}^{m}(P \cup P_j)) = m$$

means that m tokens are distributed in m source and sink places while all the other places are empty. Since very pair of i_j and o_j are strongly connected in the trivial extension of N_j, o_j must have one and only one token at M' if i_j has no token at M'. Starting from an arbitrary reachable marking, there exists such a final state; hence, the first requirement of Definition 4.7 ensures that a basic workflow can terminate properly once it participates in a run and no token is left in channel places.

Example 4.3 Figure 4.6 models a very simple version of an inter-organisational workflow that involves four organisations: client, store, bank, and logistics company. Obviously, this design is not compatible because marking $\{i_3, i_4, o_1, o_2\}$ is reachable but $\{o_1, o_2, o_3, o_4\}$ cannot be reached from it. Marking $\{i_3, i_4, o_1, o_2\}$ means that a client sends an order and then cancels it, which is reasonable obviously. In this case, bank and logistics are not involved. Obviously, collaborative-ness can reflect this kind of design requirement but compatibility cannot.

Theorem 4.6 ([2]) *If an inter-organisational workflow is compatible, then it is collaborative; but not vice versa.*

Similarly, *weak collaborative-ness* can be defined when the second requirement in Definition 4.7 is canceled. Weak collaborative-ness or collaborative-ness can ensure both deadlock-free and livelock-free.[2]

Given an inter-organisational workflow net $N^{\otimes} = (N_1, \ldots, N_m, P, F)$, a reachable marking $M \in R(N^{\otimes}, \{i_1, \ldots, i_m\})$ is a *deadlock* if the following holds:

$$\left(\exists j \in \mathbb{N}_m^+ : M(i_j) = M(o_j) = 0\right) \wedge \left(\forall t \in \bigcup_{j=1}^{m} T_j : \neg M[t\rangle\right).$$

A reachable marking $M \in R(N^{\otimes}, \{i_1, \ldots, i_m\})$ is a *livelock* if the following holds:

$$\left(\forall M' \in R(N^{\otimes}, M), \exists t \in \bigcup_{j=1}^{m} T_j : M'[t\rangle\right) \wedge \neg\left(M'\left(\bigcup_{j=1}^{m}\{i_j, o_j\}\right) = M'\left(\bigcup_{j=1}^{m}(P \cup P_j)\right) = m\right).$$

Here, a deadlock means that some basic workflows have not reached their final states after they start, but no action can be executed any more. A livelock means such a situation at which some basic workflows have been started but cannot reach their final states, while some actions can still be executed.

The following conclusion obviously holds:

Theorem 4.7 ([2]) *if an inter-organisational workflow net is collaborative, then it has neither deadlock nor livelock.*

Using the reachability graph technique or the unfolding technique can check (weak) soundness, (weak) compatibility, and (weak) collaborative-ness.

4.3   Resource Allocation Nets

This section introduces another subclass of Petri nets, named as *resource allocation nets*. The model of the dining philosophers problem belongs to this subclass. Such a Petri net generally contains a group of processes that share finitely many resources but do not consume any resource. After a process obtains one or more resources, it will use them to perform a task. After the task is done, these resources are released so that the other processes can use them.

Definition 4.8 (*Resource Allocation Net*) A net $N^{\oplus} \triangleq (N_1, \ldots, N_m, P, F)$ is called a *resource allocation net* if

---

[2] In theory, if an inter-organisational workflow net has no any enabled transition at its initial marking, it also satisfies the definition of weak collaborative-ness. However, we think nobody can design such a workflow and this case is easily checked.

1. $N_1 = (P_1, T_1, F_1), \ldots$, and $N_m = (P_m, T_m, F_m)$ are pairwise disjoint workflow nets, called *basic workflow nets*, such that the jth N_j is k_j-sound where k_j is a positive integer, $j \in \mathbb{N}_m^+$ and $m \geq 1$,
2. P is a finite set of *resource places* such that $P \cap P_j = \emptyset$ for each $j \in \mathbb{N}_m^+$,
3. $F \subseteq \bigcup_{j=1}^m (P \times T_j) \cup (T_j \times P)$ is a set of arcs representing the allocation and release of resources, and
4. for each resource place $p \in P$, there exists a minimal P-invariant X such that $\|X\| \cap P = \{p\}$.

> The fourth condition means the preservation of every resource, i.e., a resource allocated to a process should be released eventually by this process, and every released resource have previously been occupied.

Generally, a *reasonable initial marking* of a resource allocation net should satisfy that the source place of the j-th basic workflow net N_j has k_j tokens since N_j is k_j-sound, and every resource place has at least one token. For convenience of analysis, each pair of source and sink places are combined together (e.g. the places $p_{j,1}$ in the Petri net in Fig. 1.1 where $j \in \mathbb{N}_5^+$), or each workflow net becomes its trivial extension, or all source and sink places are deleted from it (e.g. the Petri net in Fig. 3.10).

We let M_0 be a reasonable initial marking of a resource allocation net and M_d be the *final marking* such that the sink place o_j has k_j tokens and the number of tokens in each resource place is equal to its initial case. Note that since the basic workflow net N_j is k_j-sound, all places in N_j except the sink one have no token at M_d. In other words, M_d means that each workflow net terminates properly and each resource restores to its initial state.

> Due to the preservation of each resource, we know that each resource restores to its initial state when each basic workflow net terminates properly. Additionally, due to the preservation of each resource and the soundness of each basic workflow net, we know that a resource allocation net is bounded at any reasonable initial marking.

If there is a reachable marking (say M) from which the final marking cannot be reached forever, there is a group of basic workflow nets that cannot terminate properly from M. Because each basic workflow net can always terminate properly when resources are not considered, we can derive that the reason causing this non-proper-termination at M is that they are occupying and waiting for some resources. Therefore, these resource places form a siphon unmarked at M. Therefore, we have the following conclusion:

Theorem 4.8 *Given a resource allocation net $N^{\oplus}$ and a reasonable initial marking M_0, the final marking M_d is reachable from each marking $M \in R(N^{\oplus}, M_0)$ if and only if each siphon containing at least one resource place is marked at each reachable marking.*

In a resource allocation system, (partial) deadlocks easily occur since some circular waits of occupying and waiting for finitely many resources take place easily. Therefore, a number of deadlock prevention or avoidance policies are proposed on the basis of Petri nets [20–31]. The reason causing deadlocks lies in some siphons that can become empty in a run, and thus most of these policies are to avoid the occurrence of an empty siphon.

Example 4.4 As shown in Fig. 2.3 in Chap. 2, the deadlock of the Petri net modelling the dining philosophers problem is caused by the siphon

$$\{p_{j,3}, r_j \mid j \in \mathbb{N}_5^+\},$$

since this siphon can be emptied by transitions $t_{j,1}$ ($j \in \mathbb{N}_5^+$). These transitions only consume tokens from this siphon but does not input any one into it. We may take the policy proposed in [20] to prevent the occurrence of such a deadlock for this system. This policy augments a control place for every such siphon. First, the number of the initial tokens in a control place is less than the total number of the initial tokens in the controlled siphon, so that the controlled siphon still has at least one token when the control place becomes empty. Second, if there is a transition (say t) only consumes a token from the controlled siphon but dose not input any token into it, then a control arc is constructed from this control place to the *source transition* of the workflow net to which t belongs. For example, $t_{1,1}$ only consumes a token from this siphon but does not produce any token into it, and coincidentally $t_{1,1}$ is also the source transition of this workflow net. The purpose of such an arc controlling a source transition is to avoid that the augmented places form a new emptied siphon. In addition, another transition is selected from this workflow net and another arc is constructed from this transition to this control place. The purpose of this arc is to return a token to the control place. Based on this policy, we can get the controlled model as shown in Fig. 4.7 which is still a resource allocation net. This controlled model is live, but it still has the starvation problem. Most of these deadlock prevention policies cannot solve the starvation problem.

When a new constraint that each basic workflow net is acyclic is added in a resource allocation net, then the resource allocation net becomes a G-system defined by Zouari and Barkaoui [19][3]; furthermore, if each basic workflow net is an acyclic state machine and each step uses one and only one resource, then it becomes an S^3PR defined by Ezpeleta et al. [20].

In theory, the deadlock problem of resource allocation systems is PSPACE-complete [32]; and even for the simplest version (i.e., each process is linear such

---

[3] A G-system is actually a weighted Petri net, i.e., each step of a process can use multiple unites of a type of resource. In this book we do not consider weighted Petri nets.

Fig. 4.7 The controlled model of the Petri net in Fig. 1.1. A control place c is augmented into it to prevent the siphon $\{p_{j,3}, r_j \mid j \in \mathbb{N}_5^+\}$ be emptied. The number of the initial tokens in c is 4 since this siphon has 5 tokens initially. This initial number may be 2 since at most two philosophers can eat simultaneously

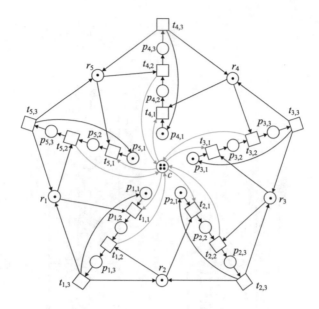

that there is neither choice, nor loop, nor parallel, and each step only uses one kind of resource), it is still NP-complete [28, 32].

Question

Try to provide the proofs of these complexity conclusions.

4.4 Application

In many real applications, multiple parallel processes both share some common resources and have to interact and collaborate via sending and receiving messages, making the related models more complex and beyond the scope of inter-organisational workflow nets and resource allocation nets. The model of solving the starvation problem in Fig. 2.4 is such an example. Here, another example of elevator scheduling system [2] is used to illustrate such a more complex model.

Scheduling

A television tower has two elevators that shuttle visitors up to the visitor platform from the ground or down to the ground from the visitor platform. They work through a scheduling system. This system may be viewed as a composition of six control flow: a *central control flow* (CCF) as shown in the centre in Fig. 4.8, a *down control flow* (DCF) as shown at the top in Fig. 4.8, an *up control flow* (UCF) as shown at the bottom in Fig. 4.8, a *left elevator control flow* (LECF) as shown on the left in

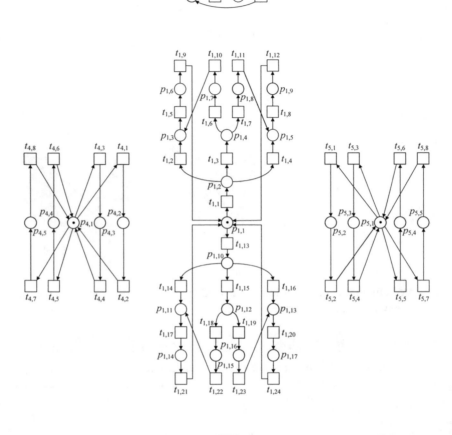

Fig. 4.8 Six basic control flows of an elevator scheduling system

Fig. 4.8, and a *right elevator control flow* (RECF) as shown on the right in Fig. 4.8. Figure 4.9 shows this system in which the six control flows are combined by places r_1-r_4 and c_1-c_{20}. Places r_1-r_4 represents the positions where the two elevators stop, which may be viewed as resource places. Places c_1-c_{20} may be viewed as channels transmitting the signals among the control flows.

CCF may receive ($t_{1,1}$ or $t_{1,13}$) a requirement of either needing an elevator down to the ground (a token in c_1 in Fig. 4.9 represents this requirement) or needing an elevator up to the platform (a token in c_3 in Fig. 4.9 represents it). Next, we only describe how to handle the down requirement because handling the up requirement

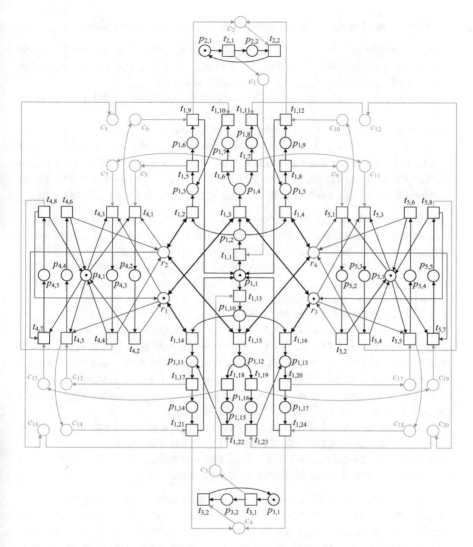

Fig. 4.9 A live scheduling design in which a starvation possibly happens. A self-loop is represented by an arc with two arrowheads at both ends

is the same as it. After CCF receiving a down requirement (i.e. firing $t_{1,1}$ in Fig. 4.9), it checks which elevator stops at the platform. There are five cases:

- the left elevator is now available, i.e., $t_{1,2}$ is enabled, or
- the right one is now available, i.e., $t_{1,4}$ is enabled, or
- the two elevators both stop at the ground, i.e., $t_{1,3}$ is enabled, or
- one is running and another one stops at the ground, or
- they are both running.

For the above five cases, the following three points are worthy noting:

- First, the token in r_2 (respectively r_4) represents that the left (respectively right) elevator stops at the platform; similarly, the token in r_1 (respectively r_3) represents that the left (respectively right) elevator stops at the ground.
- – Second, if the two elevators are both available, then CCF may randomly schedule one in this model (i.e., free choice in Petri nets).
- Last, in order to simplify the model we let CCF be in a wait state when the last two cases take place. In fact, when CCF detects that an elevator is running down to the ground and another one stops at the ground, it may schedule the elevator at the ground up to the platform such that the efficiency is enhanced. But for the simplification of this model we let CCF be in a wait state for the last two cases.

If CCF detects that one elevator now stops at the platform (without loss of generalisation, this elevator is assumed to be the left one), it enters a state ($p_{1,3}$) in which it prepares to send LECF a requirement of opening its door. Then, CCF sends LECF the requirement of opening its door (c_9). After receiving the acknowledgement that the door has been closed (c_6), CCF sends DCF an acknowledgement (c_2) that allows DCF to produce a new down requirement and thus finishes this scheduling.

If CCF detects that the two elevators both stop at the ground, then it decides to schedule one up to the platform. CCF also chooses one randomly in this model. Without loss of generalisation, let CCF choose the left one, i.e., it sends LECF a requirement of moving up (c_7). After receiving an acknowledgement that the left elevator has arrived at the platform, CCF also enters into the state in which it prepares to send LECF a requirement of opening its door ($t_{1,10}$, $p_{1,3}$, c_8). What happens later is the same with the description in the above paragraph.

As for DCF and UCF, the condition permitting a new requirement is that the previous requirement must be executed, i.e., c_2 or c_4 has a token.

Next, we describe the execution of LECF, which is similar to RECF's. At its initial state, there are the following four possible executions.

- LECF receives from CCF a requirement of opening its door when the elevator stops at the platform. After users enter the elevator and its door is closed, LECF sends CCF an acknowledgement that the door has been closed. These events are modelled by $t_{4,1}$, c_5, and c_6. The token in $p_{4,2}$ represents that this elevator is running down to the ground. Firing $t_{4,2}$ means that the elevator arrives at the ground.
- LECF receives from CCF a requirement of moving up when the elevator stops at the ground. At this time, it does not open the door but directly commands the elevator to move up. After the elevator arrives at the platform, it notifies CCF. $t_{4,3}$, $t_{4,4}$, $p_{4,3}$, c_7, and c_8 model this execution.
- LECF receives from CCF a requirement of opening its door when the elevator stops at the ground. This execution is similar to the first case (but the direction is opposite) and modelled by $t_{4,5}$, $t_{4,6}$, $p_{4,4}$, c_{13}, and c_{14}.
- LECF receives from CCF a requirement of moving down when the elevator stops at the platform. This execution is similar to the second case (also, the direction is opposite) and modelled by $t_{4,7}$, $t_{4,8}$, $p_{4,5}$, c_{15}, and c_{16}.

Control for Fairness

The Petri net in Fig. 4.9 is live and bounded, but it has a starvation (i.e., unfair). For example, after UCF sends an up requirement, CCF has not responded to this requirement (although it always has chances to do it). CCF always goes to respond to these down requirements that are infinitely sent by DCF. Formally, the following infinite transition sequence describes this case:

$$t_{3,1}\left(t_{2,1}t_{1,1}t_{1,3}t_{1,6}t_{4,3}t_{4,4}t_{1,10}t_{1,5}t_{4,1}t_{4,2}t_{1,9}t_{2,2}\right)^{*}.$$

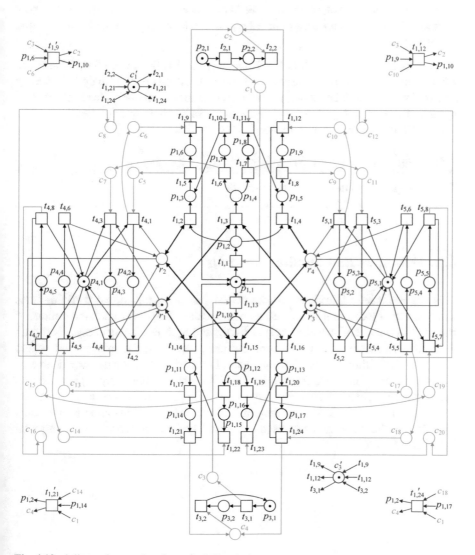

Fig. 4.10 A live and starvation-free scheduling design

To solve the starvation problem, there are two alternative methods.

– One is that when CCF finishes an up (respectively down) requirement, it first checks if there has been a down (respectively up) requirement. If so, then it continues to respond to this requirement. If not, it directly returns to its initial state. Figure 4.10 shows the model in which the starvation problem is solved by this method.
– Another method is that when an elevator process responds to a requirement of moving up (respectively down), it first goes to check if there has been an up (respectively down) requirement. If so, it goes to open the door and responds to this requirement. If not, it directly commands the elevator to move up (respectively down) and does not open its door. Here, the model corresponding to this method is omitted.

Two tokens may be put into the place $p_{1,1}$ so that the scheduling system can advance the utilisation rate of elevators (because CCF may handle an up requirement and an down one simultaneously). However, deadlocks can take place in the related system. To solve this problem, some mutually exclusive measures must be taken when CCF handles two requirements at the same time.

Question

Try to design such a deadlock-free controller.

4.5   Summery and Further Reading

This chapter briefly introduces workflow nets, inter-organisational workflow nets and resource allocation nets as well as some related properties such as soundness, compatibility and collaborative-ness. Mining and optimising workflows (based on Petri nets and machine learning) from log files of an enterprise information system are some interesting studies [33, 34]. How to calculate the consistency degree of two workflows is an interesting topic [35–37]. Another interesting topic is how to model the data operations in workflows so that some flaws such as data inconsistency can be checked [38–43]. Another important study is how to schedule resources in a workflow so that they can be utilised efficiently, and an important way is based on Petri nets [44, 45].

References

1. van der Aalst, W.M.P.: Structural Characterizations of Sound Workflow Nets. *Computing Science Report* 96/23, Eindhoven University of Technology, pp. 1–23 (1996)
2. Liu, G.J., Zhou, M.C., Jiang, C.J.: Petri net modelling and Collaborativeness for Parallel Processes with Resource Sharing and Message Passing. *ACM Transactions on Embedded Computing Systems* 16 (article no. 113): 1-20 (2017)
3. Liu, G.J., Liu, Y., Sun, J., Dong, J.S.: Complexity of the soundness problem of workflow nets. *Fundamenta Informaticae* 131: 81–101, (2014)
4. Tiplea, F.L., Bocaenealae, C., Chirosecae, R.: On the complexity of deciding soundness of acyclic workflow nets. *IEEE Transactions on Systems, Man and Cybernetics: Systems* 45: 1292–1298 (2015)
5. Liu, G.J.: PSPACE-completeness of the soundness problem of safe asymmetric-choice workflow nets. In: *the 41st International Conference on Application and Theory of Petri Nets and Concurrency*, pp. 196–216 (2020)
6. Liu, G.J.: Some complexity results for the soundness problem of workflow nets. *IEEE Transactions on Services Computing* 7: 322–328 (2014)
7. Liu, G.J., Jiang, C.J.: Co-NP-hardness of the soundness problem for asymmetric-choice workflow nets. *IEEE Transactions on Systems, Man and Cybernetics: Systems* 45: 1201–1204 (2015)
8. Desel, J., Esparza, J.: *Free Choice Petri Nets*. Cambridge University Press (1995)
9. Hee, K.V., Sidorova, N., Voorhoeve, M.: Soundness and separability of workflow nets in the stepwise refinement approach. In: *the 24th International Conference on Applications and Theory of Petri Nets*, pp. 337–356 (2003)
10. van der Aalst, W.M.P.: Interorganizational workflows: an approach based on message sequence charts and Petri nets. *Systems Analysis Modelling Simulation* 34: 335–367 (1999)
11. Martens, A.: On compatibility of web services. *Petri Net Newsletters* 65: 12–20 (2003)
12. van der Aalst, W.M.P., Hee, K.V., Hofstede, A., Sidorova, N., Wynn, M.T.: Soundness of workflow nets: classification, decidability, and analysis. *Formal Aspects of Computing* 23: 333–363 (2011)
13. Liu, G.J., Jiang, C.J., Zhou, M.C.: Process nets with channels. *IEEE Transactions on Systems, Man and Cybernetics - Part A: Systems and humans* 42: 213–225 (2012)
14. Liu, G.J., Jiang, C.J., Zhou, M.C., Xiong, P.C.: Interactive Petri nets. *IEEE Transactions on Systems, Man and Cybernetics: Systems* 43: 291–302 (2013)
15. Liu, G.J., Chen, L.J.: Sufficient and necessary condition for compatibility of a class of interorganizational workflow nets. *Mathematical Problems in Engineering* 2015 (article no. 392945): 1–11 (2015)
16. Liu, G.J., Jiang, C.J.: Net-structure-based conditions to decide compatibility and weak compatibility for a class of inter-organizational workflow nets. *SCIENCE CHINA Information Science* 58 (article no. 072103): 1–16 (2015)
17. Liu, G.J., Reisig, W., Jiang, C.J., Zhou, M.C.: A branching-process-based method to check soundness of workflow systems. *IEEE Access* 4: 4104–4118 (2016)
18. Liu, G.J., Zhang, K., Jiang, C.J.: Deciding the deadlock and livelock in a Petri net with a target marking based on Its basic unfolding. In: *the 16the IEEE International Conference on Algorithms and Architectures for Parallel Processing*, pp. 98–105 (2016)
19. Zouari, B., Barkaoui, K.: Parameterized supervisor synthesis for a modular class of discrete event systems. In: *IEEE International Conference on Systems, Man and Cybernetics*, pp. 1874–1879 (2003)
20. Ezpeleta, J., Colom, J.M., Martinez, J.: A petri net based deadlock prevention policy for flexible manufacturing systems. *IEEE Transactions on Robotics and Automation* 11: 173–184 (1995)
21. Liu, G.J., Jiang, C.J., Wu, Z.H., Chen, L.J.: A live subclass of Petri nets and their application in modeling flexible manufacturing systems. *International Journal of Advanced Manufacturing Technology* 41: 66–74 (2009)

22. Liu, G.J., Jiang, C.J., Chen, L.J., Wu, Z.H.: Two types of extended RSNBs and their application in modelling flexible manufacturing systems. *International Journal of Advanced Manufacturing Technology* 45: 573–582 (2009)
23. Liu, G.J., Jiang, C.J., Zhou, M.C.: Two simple deadlock prevention policies for S^3PR based on key-resource/operation-place pairs. *IEEE Transactions on Automation Science and Engineering* 7: 945–957 (2010)
24. Liu, G.J., Jiang, C.J., Zhou, M.C.: Improved condition for controllability of strongly dependent strict minimal siphons in Petri nets. In: *the 7th International Conference on Networking, Sensor and Cybernetics*, pp. 359–364 (2011)
25. Liu, G.J., Jiang, C.J., Zhou, M.C.: Improved sufficient condition for the controllability of weakly dependent siphons in system of simple sequential processes with resources. *IET Control Theory and Applications* 5: 1059–1068 (2011)
26. Liu, G.J., Jiang, C.J., Zhou, M.C., Ohta, A.: The liveness of WS^3PR: complexity and decision. *IEICE Transactions on Fundamentals of Electronics, Communications and Computer Sciences* E96-A: 1783–1793 (2013)
27. Chao, D.Y., Liu, G.J.: A simple suboptimal siphon-based control model of a well-known S^3PR. *Asian Journal of Control* 14: 163–172 (2012)
28. Reveliotis, S.A.: *Real-time Management of Resource Allocation Systems: A Discrete Event Systems Approach*. Springer, New York (2005)
29. Li, Z.W., Zhou, M.C.: *Deadlock Resolution in Automated Manufacturing Systems: A Novel Petri Net Approach*. New York: Springer (2009)
30. Khaleghi, R., Sreenivas, R.S.: On computing the supremal right-closed control invariant subset of a right-closed set of markings for an arbitrary petri net. *Discrete Event Dynamic Systems* 31: 373–405 (2021)
31. Boucheneb, H., Barkaoui, K., Xing, Q., Wang, K., Liu, G., Li, Z.W.: Time based deadlock prevention for Petri nets. *Automatica* 137 (article no. 110119): 1–11 (2022)
32. Liu, G.J.: Complexity of the deadlock problem for Petri nets modeling resource allocation systems. *Information Sciences* 363: 190–197 (2016)
33. van der Aalst, W.M.P.: Using free-choice nets for process mining and business process management. In: *the 16th Conference on Computer Science and Intelligence Systems*, pp. 9–15 (2021)
34. Lima, F., Cortez, M.F., Schmidt, P.P., Silverio, A.K., da Silva Fernandes, A.M., Junior, J.C.: Simulation of the ambulatory processes in the biggest Brazilian cardiology hospital: a Petri net approach. *Independent Journal of Management and Production* 12: 219–240 (2021)
35. Wang, M., Liu, G.J., Yan, C., Jiang, C.J.: Behavior consistency computation for workflow nets with unknown correspondence. *IEEE/CAA Journal of Automatica Sinica* 5: 281–291 (2018)
36. Wang, M., Ding, Z.J., Liu, G.J., Jiang, C.J., Zhou, M.C.: Measurement and computation of profile similarity of workflow nets based on behavioral relation matrix. *IEEE Transactions on Systems, Man, and Cybernetics: Systems* 50: 3628–3645 (2020)
37. Zhao, F., Xiang, D.M., Liu, G.J., Jiang, C.J.: A new method for measuring the behavioral consistency degree of WF-Net systems. *IEEE Transactions on Computational Social Systems* 9: 480–493 (2022)
38. Xiang, D.M., Liu, G.J., Yan, C.G., Jiang, C.J.: Detecting data inconsistency based on the unfolding technique of Petri nets. *IEEE Transactions on Industrial Informatics* 13: 2995–3005 (2017)
39. Xiang, D.M., Liu, G.J., Yan, C.G., Jiang, C.J.: Detecting data-flow errors based on Petri nets with data operations. *IEEE/CAA Journal of Automatica Sinica* 5: 251–260 (2018)
40. Xiang, D.M., Liu, G.J.: Checking data-flow errors based on the guard-driven reachability graph of WFD-Net. *Computing and Informatics* 39: 1001–1020 (2020)
41. Xiang, D.M., Liu, G.J., Yan, C.G., Jiang, C.J.: A guard-driven analysis approach of workflow net with data. *IEEE Transactions on Services Computing* 14: 1675–1686 (2021)
42. Xiang, D.M., Lin, S., Wang, X.H., Liu, G.J.: Checking missing-data errors in cyber-physical systems based on the merged process of Petri nets. *IEEE Transactions on Industrial Informatics* (It has been published online) (2022)

43. Tao, X.Y., Liu, G.J., Yang, B., Yan, C., Jiang, C.J.: Workflow nets with tables and their soundness. *IEEE Transactions on Industrial Informatics* 16: 1503–1515 (2020)
44. Viriyasitavat, W., Xu, L.D., Dhiman, G., Sapsomboon, A., Pungpapong, V., Bi, Z.: Service workflow: state-of-the-art and future trends. *IEEE Transactions on Services Computing* (2021)
45. Li, H., Wang, B., Yuan, Y., Zhou, M.C., Fan, Y.S., Xia, Y.: Scoring and dynamic hierarchy-based NSGA-II for multi-objective workflow scheduling in the cloud. *IEEE Transactions on Automation Science and Engineering* (2021)

Chapter 5
Verifying Computation Tree Logic Based on Petri Nets

5.1 Computation Tree Logic and Verification Based on Petri Nets

This section first presents the syntax and semantics of computation tree logic (CTL) and then describes the general algorithms of verifying CTL formulae on the basis of the reachability graphs of Petri nets. CTL was independently defined by Emerson and Clarke [1] and Queille and Sifakis [2] for model checking, and then have widely been studied and applied [3, 4]. The following syntax and semantics of CTL refer to the work in [5].

5.1.1 Syntax and Semantics of CTL

5.1.1.1 Syntax of CTL

A CTL formula describes a property of a *computation tree* and is generally composed of *path quantifiers* and *temporal operators*.

The path quantifiers include **A** and **E** representing "for all computation paths" and "for some computation path", respectively. When they are used in a state, it means that all of the paths or some of the paths starting from the state have some property.

The temporal operators include **X**, **F**, **G** and **U**, describing properties of a path:

- the operator **X** ("next") requires that a property holds in the second state of this path,
- the operator **F** ("in the future" or "eventually") requires that a property will hold at some state of this path,
- the operator **G** ("always" or "globally") asserts that a property holds at every state of this path, and
- the operator **U** describes two properties of this path such that the first one always holds along with the path until the second one holds at some state of the path.

G. Liu, *Petri Nets*,
https://doi.org/10.1007/978-981-19-6309-4_5

Definition 5.1 (*Syntax of CTL*) Let AP be the set of atomic propositions. Then, CTL formulae can be defined according to the following grammar:

$$\phi := \textbf{true} \mid \textbf{false} \mid \textbf{deadlock} \mid p \mid \neg\phi \mid \phi_1 \vee \phi_2 \mid \phi_1 \wedge \phi_2 \mid \phi_1 \Rightarrow \phi_2 \mid$$

$$\textbf{EX}\ \phi \mid \textbf{EF}\ \phi \mid \textbf{EG}\ \phi \mid \textbf{E}(\phi_1\ \textbf{U}\ \phi_2) \mid \textbf{AX}\ \phi \mid \textbf{AF}\ \phi \mid \textbf{AG}\ \phi \mid \textbf{A}(\phi_1\ \textbf{U}\ \phi_2)$$

where $p \in AP$, and **true**, **false**, and **deadlock** are constants that are not in AP.

5.1.1.2 Labelled Transition System and Computation Path

Formally explaining CTL semantics needs a so-called *labelled transition system* defined as follows:

Definition 5.2 (*Labelled Transition System*) A *labelled transition system* is 6-tuple $\mathbb{A} \triangleq (St, Act, \longrightarrow, s_0, AP, L)$ where

1. St is a set of *states*,
2. Act is a set of *actions*,
3. $\longrightarrow \subseteq (St \times Act \times St)$ is a *transition relation*,
4. $s_0 \in St$ is the *initial state*,
5. AP is the set of *atomic propositions*, and
6. $L: St \rightarrow 2^{AP}$ is a label function.

In fact, a labelled transition system is an automaton $(St, Act, \longrightarrow, s_0)$ in which each state is labelled by a group of atomic propositions. A state s is called *terminal* if the following holds:

$$\forall s' \in St, \forall a \in Act : (s, a, s') \notin \longrightarrow .$$

i.e., it has no succeeding state (successor). Instead of the form $(s, a, s') \in \longrightarrow$, we write it as $s \xrightarrow{a} s'$. Since actions are not used in the semantics specification and verification of CTL formulae, we omit the action name and use the form $s \longrightarrow s'$ when writing a transition.

> If an atomic proposition is associated with a state, it means that this proposition holds when the automata is at the state.

Formally speaking, for each state $s \in St$ and for each proposition $p \in AP$, p *holds* (is true) at s if and only if

$$p \in L(s).$$

Notice that Boolean constants **true** (which is thought of as the proposition that is always true) and **false** (which is thought of as the proposition that is false no matter what) are not in AP, and **deadlock** (which means a terminal state) is viewed as a special CTL formula. Boolean constants **true** and **false** usually are represented by 1 and 0, respectively.

A *computation path* starting from some state (say s_1) in a labelled transition system is either a *maximal finite state sequence*

$$\omega = s_1 \cdots s_k$$

such that $k \geq 1$, s_k is a terminal state and for each $j \in \mathbb{N}_{k-1}^+$ there is

$$s_j \longrightarrow s_{j+1},$$

or an *infinite state sequence*

$$\omega = s_1 s_2 \cdots$$

such that for each $j \in \mathbb{N}^+$ there is

$$s_j \longrightarrow s_{j+1}.$$

For the first case, the *position set* of the computation path is defined as

$$positions(\omega) = \mathbb{N}_k^+,$$

and for the second one its *position set* is defined as

$$positions(\omega) = \mathbb{N}^+.$$

Given a computation path ω and a position $j \in positions(\omega)$, $\omega(j)$ represents the j-th state of this path. Given a state s, $\Omega(s)$ represents the set of all computation paths starting from s.

Example 5.1 For instance of the labelled transition system in Fig. 5.1a, it has 6 states s_j ($j \in \mathbb{N}_5$) in which s_0 is the initial state and s_5 is a terminal one. It has 8 atomic propositions r_1, r_2 and $p_{j,k}$ associated with these states, where $j \in \mathbb{N}_2^+$ and $k \in \mathbb{N}_3^+$. Note that a finite computation path should be viewed as a maximal state sequence in the unfolded tree of a labeled transition system, as shown in Fig. 5.1b including

$$s_0 s_1 s_5, \quad s_0 s_3 s_5, \quad s_0 s_1 s_2 s_0 s_1 s_5, \quad s_0 s_1 s_2 s_0 s_3 s_5, \quad \ldots.$$

Example 5.2 For the labelled transition system in Fig. 5.1, we can construct the following formulae that will be explained and verified later:

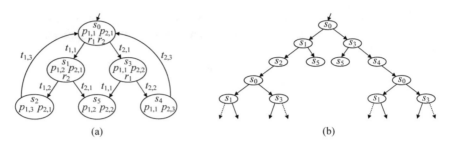

Fig. 5.1 **a** A labelled transition system with a terminal state, and **b** its computation tree generated via unfolding loops

- $\phi_1 = \mathbf{EF}(\mathbf{AG}(\neg r_1 \wedge \neg r_2))$,
- $\phi_2 = \mathbf{EG}(p_{1,2} \Rightarrow \mathbf{AF}p_{1,1})$.

5.1.1.3   Semantics of CTL

Given a labelled transition system $\mathbb{A} = (St, Act, \longrightarrow, s_0, AP, L)$, a state $s \in St$ and a CTL formula ϕ, the notation $(\mathbb{A}, s) \models \phi$ denotes that the formula ϕ holds at the state s (alternatively, s satisfies ϕ, ϕ is satisfiable at s, or s is a satisfiable state of ϕ), and it is simplified into $s \models \phi$ via omitting $\mathbb{A}$ when no ambiguity takes place.

Definition 5.3 (*Semantics of CTL*) The relation $\models$ is defined inductively as follows:

- $s \models \mathbf{true}$,
- $s \models \mathbf{false}$ if and only if $s \not\models \mathbf{true}$,
- $s \models \mathbf{deadlock}$ if and only if s is a terminal state,
- $s \models p$ if and only if $p \in L(s)$,
- $s \models \neg\phi$ if and only if ϕ does not hold at state s (which is denoted as $s \not\models \phi$),
- $s \models \phi_1 \vee \phi_2$ if and only if $s \models \phi_1$ or $s \models \phi_2$,
- $s \models \phi_1 \wedge \phi_2$ if and only if $s \models \phi_1$ and $s \models \phi_2$,
- $s \models \phi_1 \Rightarrow \phi_2$ if and only if $s \not\models \phi_1 \vee s \models \phi_2$,
- $s \models \mathbf{EX}\,\phi$ if and only if there exists $s' \in St$ such that $s \longrightarrow s'$ and $s' \models \phi$,
- $s \models \mathbf{AX}\,\phi$ if and only if for each $s' \in St$, if $s \longrightarrow s'$ then $s' \models \phi$,
- $s \models \mathbf{EF}\,\phi$ if and only if there exists $\omega \in \Omega(s)$ and $j \in positions(\omega)$ such that $\omega(j) \models \phi$,
- $s \models \mathbf{AF}\,\phi$ if and only if for each $\omega \in \Omega(s)$ there exists $j \in positions(\omega)$ such that $\omega(j) \models \phi$,
- $s \models \mathbf{EG}\,\phi$ if and only if there exists $\omega \in \Omega(s)$ such that for each $j \in positions(\omega)$ there is $\omega(j) \models \phi$,
- $s \models \mathbf{AG}\,\phi$ if and only if for each $\omega \in \Omega(s)$ and for each $j \in positions(\omega)$ there is $\omega(j) \models \phi$,
- $s \models \mathbf{E}(\phi_1\,\mathbf{U}\,\phi_2)$ if and only if there exists $\omega \in \Omega(s)$ and $j \in positions(\omega)$ such that $\omega(j) \models \phi_2$ and for each $k \in \mathbb{N}_{j-1}^+$ there is $\omega(k) \models \phi_1$,

Fig. 5.2 The Petri net modelling the dining 2-philosophers' problem. Its reachability graph is thought of as a labelled transition system as shown in Fig. 5.1

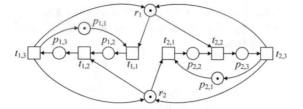

$- s \models \mathbf{A}(\phi_1 \; \mathbf{U} \; \phi_2)$ if and only if for each $\omega \in \Omega(s)$ there exists $j \in positions(\omega)$ such that $\omega(j) \models \phi_2$ and for each $k \in \mathbb{N}^+_{j-1}$ there is $\omega(k) \models \phi_1$.

> Existing a successor of a given state is a necessary condition on ensuring $\mathbf{EX} \; \phi$ hold at this state.

Before explaining the formulae in Example 5.2, we first introduce the background of the labelled transition system in Fig. 5.1. In fact, it is the reachability graph of the Petri net in Fig. 5.2 that models the dining 2-philosophers' problem, and we can have the following reasonable assumptions:

> Each place of the Petri net is viewed as an atomic proposition, each reachable marking is viewed as a state, and an atomic proposition is true at a state if and only if the corresponding place is marked at the corresponding marking.

Example 5.3 The formula

$$\phi_1 = \mathbf{EF}(\mathbf{AG}(\neg r_1 \wedge \neg r_2))$$

means that there exists a state such that neither r_1 nor r_2 is marked at all the states reachable from this one. The sub-formula

$$\mathbf{AG}(\neg r_1 \wedge \neg r_2)$$

holds at the state s_5 but does not hold at others. Obviously, $s_j \models \phi_1$ for each $j \in \mathbb{N}_5$. This formula shows that, in the Petri net in Fig. 5.2, there is such a run during which all resources have been allocated but cannot been released any more, i.e., deadlock.

Example 5.4 The formula

$$\phi_2 = \mathbf{EG}(p_{1,2} \Rightarrow \mathbf{AF}p_{1,1}) = \mathbf{EG}(\neg p_{1,2} \vee \mathbf{AF}p_{1,1})$$

means that there exists a computation path such that, for each state of this computation path, if $p_{1,2}$ is marked at this state then $p_{1,1}$ can always be marked in the future. The sub-formula

$$p_{1,2} \Rightarrow \mathbf{AF}\, p_{1,1}$$

holds at states s_0, s_2, s_3, and s_4, since the precondition $p_{1,2}$ of this implication expression does not hold at them at all; but it does not hold at the state s_1 (respectively the state s_5) since the precondition $p_{1,2}$ holds at it but there exists a computation path $s_1 s_5$ (respectively a computation path s_5) that does not satisfy $\mathbf{AF}(p_{1,1})$. Although the sub-formula holds at states s_0, s_2, s_3 and s_4, the formula ϕ_2 does not hold at any of them because, from any of them, there is a computation path containing s_5.

Example 5.5 If we re-write the formula ϕ_2 as

$$\phi_3 = \mathbf{EG}(p_{1,2} \Rightarrow \mathbf{EF}\, p_{1,1}) = \mathbf{EG}(\neg p_{1,2} \vee \mathbf{EF}\, p_{1,1}),$$

then ϕ_3 holds at s_j where $j \in \mathbb{N}_4$. This formula means that when Philosopher 1 takes a chopstick, there is a run making her/him eat and restore her/his initial state.

A CTL formula ϕ is *valid* in a labelled transition system $\mathbb{A}$, denoted as $\mathbb{A} \models \phi$, if $(\mathbb{A}, s_0) \models \phi$, i.e., the formula is satisfiable at the initial state.

In the above examples, ϕ_1 and ϕ_3 are obviously valid but ϕ_2 is not.

5.1.2   Logical Equivalence of Formulae

5.1.2.1   Some Equivalent Formulae

The classical definition of CTL semantics generally requires that a labelled transition system has no terminal states for deriving the equivalence of some formulae [3, 4]. Here we take the semantics defined in [5], and thus some formulae are still equivalent when a labelled transition system has terminal states.

First, given a terminal state s, we can see the following facts:

- $s \models \mathbf{true}$,
- $s \not\models \mathbf{false}$,
- $s \models \mathbf{deadlock}$,
- $s \not\models \mathbf{EX}\, \phi$ for any ϕ,
- $s \models \mathbf{AX}\, \phi$ for any ϕ,
- $s \models \mathbf{EF}\, \phi$ if and only if $s \models \phi$,
- $s \models \mathbf{AF}\, \phi$ if and only if $s \models \phi$,
- $s \models \mathbf{EG}\, \phi$ if and only if $s \models \phi$,
- $s \models \mathbf{AG}\, \phi$ if and only if $s \models \phi$,
- $s \models \mathbf{E}(\phi_1\ \mathbf{U}\ \phi_2)$ if and only if $s \models \phi_2$,
- $s \models \mathbf{A}(\phi_1\ \mathbf{U}\ \phi_2)$ if and only if $s \models \phi_2$.

For a terminal state, there is a unique computation path, i.e., this path contains this terminal state itself only. Therefore, the above conclusions hold obviously according to Definition 5.3.

The following shows 9 equivalent transforms:

- **fasle** $= \neg$**true**.
- **deadlock** $= \neg$**EX true**.
- $\phi_1 \vee \phi_2 = \neg(\neg\phi_1 \wedge \neg\phi_2)$.
- $\phi_1 \Rightarrow \phi_2 = \neg\phi_1 \vee \phi_2$.
- **EF** $\phi =$ **E**(**true U** ϕ).
- **AX** $\phi = \neg$**EX** $\neg\phi$.
- $\mathbf{A}(\phi_1 \ \mathbf{U} \ \phi_2) = \neg\Big(\big(\mathbf{EG}\neg\phi_2\big) \vee \mathbf{E}\big(\big(\phi_1 \wedge \neg\phi_2\big)\mathbf{U}\big(\neg\phi_1 \wedge \neg\phi_2\big)\big)\Big),$
- **AF** $\phi =$ **A**(**true U** ϕ),
- **AG** $\phi = \neg$**EF** $\neg\phi$.

To understand the equivalence of **AU**, we need to know the following two cases that just make $\mathbf{A}(\phi_1 \ \mathbf{U} \ \phi_2)$ not hold at a state s:
- there exists a computation path along which ϕ_2 does not hold forever, i.e.,

$$s \models \mathbf{EG} \ \neg\phi_2,$$

- there exists a computation path such that when ϕ_1 becomes false along this path, ϕ_2 still keeps false, i.e.,

$$s \models \mathbf{E}\big(\big(\phi_1 \wedge \neg\phi_2\big)\mathbf{U}\big(\neg\phi_1 \wedge \neg\phi_2\big)\big).$$

Question

Try to provide proofs for the above equivalences.

5.1.2.2 Existential Normal Form

Based on the above equivalences, the syntax of CTL can be defined in the following *existential normal form*:

$$\phi ::= \mathbf{true} \mid p \mid \neg\phi \mid \phi_1 \wedge \phi_2 \mid \mathbf{EX} \ \phi \mid \mathbf{EG} \ \phi \mid \mathbf{E}(\phi_1 \ \mathbf{U} \ \phi_2).$$

Example 5.6 The formulae in Examples 5.3, 5.4, and 5.5 can be transferred into the existential normal forms as follows:

$-\ \phi_1 = \mathbf{EF}(\mathbf{AG}(\neg r_1 \wedge \neg r_2))$
$\qquad = \mathbf{EF}(\neg \mathbf{EF}(\neg(\neg r_1 \wedge \neg r_2)))$
$\qquad = \mathbf{EF}(\neg \mathbf{EF}(\neg(\neg r_1 \wedge \neg r_2)))$
$\qquad = \mathbf{EF}(\neg \mathbf{E}(\mathbf{true}\ \mathbf{U}\ \neg(\neg r_1 \wedge \neg r_2)))$
$\qquad = \mathbf{E}(\mathbf{true}\ \mathbf{U}\ \neg \mathbf{E}(\mathbf{true}\ \mathbf{U}\ \neg(\neg r_1 \wedge \neg r_2))),$

$-\ \phi_2 = \mathbf{EG}(p_{1,2} \Rightarrow \mathbf{AF} p_{1,1})$
$\qquad = \mathbf{EG}(\neg p_{1,2} \vee \mathbf{AF} p_{1,1})$
$\qquad = \mathbf{EG}(\neg p_{1,2} \vee \mathbf{A}(\mathbf{true}\ \mathbf{U}\ p_{1,1}))$
$\qquad = \mathbf{EG}(\neg p_{1,2} \vee \neg((\mathbf{EG}\neg p_{1,1}) \vee \mathbf{E}((\mathbf{true} \wedge \neg p_{1,1})\mathbf{U}(\neg\mathbf{true} \wedge \neg p_{1,1}))))$
$\qquad = \mathbf{EG}(\neg p_{1,2} \vee \neg((\mathbf{EG}\neg p_{1,1}) \vee \mathbf{E}(\neg p_{1,1}\ \mathbf{U}\ \mathbf{false})))$
$\qquad = \mathbf{EG}(\neg p_{1,2} \vee \neg((\mathbf{EG}\neg p_{1,1}) \vee \mathbf{false}))$
$\qquad = \mathbf{EG}(\neg p_{1,2} \vee \neg \mathbf{EG}\neg p_{1,1})$
$\qquad = \mathbf{EG}\neg(p_{1,2} \wedge \mathbf{EG}\neg p_{1,1}),$

$-\ \phi_3 = \mathbf{EG}(p_{1,2} \Rightarrow \mathbf{EF} p_{1,1})$
$\qquad = \mathbf{EG}(\neg p_{1,2} \vee \mathbf{EF} p_{1,1})$
$\qquad = \mathbf{EG}(\neg p_{1,2} \vee \mathbf{E}(\mathbf{true}\ \mathbf{U}\ p_{1,1}))$
$\qquad = \mathbf{EG}\neg(p_{1,2} \wedge \neg \mathbf{E}(\mathbf{true}\ \mathbf{U}\ p_{1,1})).$

In what follows, we only consider the verification algorithms of CTL formulae written in these existential normal forms.

5.1.3 Verification Algorithms Based on Reachability Graphs

As shown above, the reachability graph of a Petri net can be viewed as a labelled transition system. The reachable markings stand for states, the places stand for atomic propositions, and an atomic proposition is true at a state if and only if the related place is marked at the related marking.

In consequence, given the reachability graph $RG(N, M_0) = (R(N, M_0), E)$ of a Petri net $(N, M_0) = (P, T, F, M_0)$ as well as a CTL formula ϕ written in an existential normal form, we can verify if this formula is valid or not in this Petri net via computing those reachable markings that satisfy this formula. If a formula is satisfiable at the initial marking, then we say that the formula is *valid* in this Petri net.

Algorithm 5.1 describes a recursive procedure of computing the satisfiable state set of ϕ, and the computation of $\mathbf{SAT_{EX}}(\phi_1)$, $\mathbf{SAT_{EG}}(\phi_1)$, and $\mathbf{SAT_{EU}}(\phi_1, \phi_2)$ in it can be done by involving Algorithms 5.2, 5.3 and 5.4, respectively. In what follows, we explain briefly how to compute them.

To compute the satisfiable set of **EX** ϕ is exactly to find the predecessors of all the states in **SAT**(ϕ).

Algorithm 5.1: SAT(ϕ): Computing the Satisfiable Marking Set of ϕ Based on a Reachability Graph.

Input: A reachability graph $(R(N, M_0), E)$ and a CTL formula ϕ;
Output: The satisfiable set **SAT**(ϕ);

1 **if** $\phi = $ **true then**
2 $\quad$ **return** $R(N, M_0)$;

3 **if** $\phi = p$ **then**
4 $\quad$ **return** $\{M \in R(N, M_0) \mid M(p) \geq 1\}$;

5 **if** $\phi = \neg\phi_1$ **then**
6 $\quad$ **return** $\{M \in R(N, M_0) \mid M \notin \mathbf{SAT}(\phi_1)\}$;

7 **if** $\phi = \phi_1 \wedge \phi_2$ **then**
8 $\quad$ **return** $\mathbf{SAT}(\phi_1) \cap \mathbf{SAT}(\phi_2)$;

9 **if** $\phi = \mathbf{EX}\, \phi_1$ **then**
10 $\quad$ **return** $\mathbf{SAT}_{\mathbf{EX}}(\phi_1)$;

11 **if** $\phi = \mathbf{EG}\, \phi_1$ **then**
12 $\quad$ **return** $\mathbf{SAT}_{\mathbf{EG}}(\phi_1)$;

13 **if** $\phi = \mathbf{E}(\phi_1\ \mathbb{U}\ \phi_2)$ **then**
14 $\quad$ **return** $\mathbf{SAT}_{\mathbf{EU}}(\phi_1, \phi_2)$;

If a state satisfying ϕ is nonterminal and any successor of it does not satisfy ϕ, then it dose not satisfy **EG** ϕ.

To compute the satisfiable state set of **EG** ϕ, therefore, we first initialise the satisfiable set of **EG** ϕ as **SAT**(ϕ), and then delete out of this set those states that are not terminal but none of their successors is in this set. Repeating this procedure until the set has no such state, we obtain the real **SAT(EG** $\phi)$.

Each state satisfying ϕ_2 also satisfies $\mathbf{E}(\phi_1\ \mathbf{U}\ \phi_2)$.

Therefore, **SAT**(ϕ_2) is initially as the satisfiable set of $\mathbf{E}(\phi_1\ \mathbf{U}\ \phi_2)$, i.e., the second line in Algorithm 5.4. Starting from these states satisfying ϕ_2, we find their predecessors satisfying ϕ_1 and add them into the satisfiable set of $\mathbf{E}(\phi_1\ \mathbf{U}\ \phi_2)$ since they also satisfy $\mathbf{E}(\phi_1\ \mathbf{U}\ \phi_2)$. Starting from these new satisfiable states, we continue to find their predecessors satisfying ϕ_1 and add them into the satisfiable set, and so on, until there is no such predecessor.

Algorithm 5.2: SAT$_{EX}(\phi)$: Computing the Satisfiable Marking Set of **EX** ϕ Based on a Reachability Graph.

Input: A reachability graph $(R(N, M_0), E)$ and a formula ϕ;
Output: The satisfiable set **SAT$_{EX}(\phi)$**;
1 $X \leftarrow$ **SAT**(ϕ);
2 $Y \leftarrow \{M \in R(N, M_0) \mid \exists M' \in X : (M, M') \in E\}$;
3 **return** Y;

Algorithm 5.3: SAT$_{EG}(\phi)$: Computing the Satisfiable Marking Set of **EG** ϕ Based on a Reachability Graph.

Input: A reachability graph $(R(N, M_0), E)$ and a formula ϕ;
Output: The satisfiable set **SAT$_{EG}(\phi)$**;
1 $X \leftarrow$ **SAT**(ϕ);
2 $Y \leftarrow \{M \in X \mid M \not\models$ **deadlock** $\land \forall M' \in X : (M, M') \notin E\}$;
3 **while** $Y \neq \emptyset$ **do**
4 $\quad\mid\quad X \leftarrow X \setminus Y$;
5 $\quad\mid\quad Y \leftarrow \{M \in X \mid M \not\models$ **deadlock** $\land \forall M' \in X : (M, M') \notin E\}$;
6 **return** X;

Algorithm 5.4: SAT$_{EU}(\phi_1, \phi_2)$: Computing the Satisfiable Marking Set of **E$(\phi_1$ U $\phi_2)$}** Based on a Reachability Graph.

Input: A reachability graph $(R(N, M_0), E)$ and two formulae ϕ_1 and ϕ_2;
Output: The satisfiable set **SAT$_{EU}(\phi_1, \phi_2)$**;
1 $X \leftarrow$ **SAT**(ϕ_1);
2 $Y \leftarrow$ **SAT**(ϕ_2);
3 $Z \leftarrow \{M \in X \mid \exists M' \in Y : (M, M') \in E\}$;
4 **while** $Z \neq \emptyset$ **do**
5 $\quad\mid\quad Y \leftarrow Y \cup Z$;
6 $\quad\mid\quad X \leftarrow X \setminus Z$;
7 $\quad\mid\quad Z \leftarrow \{M \in X \mid \exists M' \in Y : (M, M') \in E\}$;
8 **return** Y;

Note that when we store a reachability graph, we may record not only the successors but also the predecessors for each reachable marking, so that we can fast obtain the predecessors of any given marking. But, it is also obvious that storing them in advance will occupy more storage space.

Question

Is there an effective method to compute the predecessors of a given marking if they are not stored in advance? If yes, try to provide one. Try to provide a specific time complexity for Algorithms 5.2, 5.3 and 5.4, respectively.

Example 5.7 Let's consider the computation of the satisfiable state set of the following formula based on the reachability graph in Fig. 5.1a:

- $EG(\neg(p_{1,2} \wedge \neg E(\text{true } U\ p_{1,1})))$.

First, we obtain the satisfiable sets of atomic propositions $p_{1,1}$ and $p_{1,2}$ as well as the satisfiable set of the constant **ture** as follows:

- $SAT(p_{1,1}) = \{s_0, s_3, s_4\}$,
- $SAT(p_{1,2}) = \{s_1, s_5\}$,
- $SAT(\text{true}) = \{s_0, s_1, s_2, s_3, s_4, s_5\}$.

Next, in order to understand the computation of $SAT(E(\text{true } U\ p_{1,1}))$ according to Algorithm 5.4, we illustrate the changes of those variables in Fig. 5.3 and have

- $SAT(E(\text{true } U\ p_{1,1})) = \{s_0, s_1, s_2, s_3, s_4\}$.

Consequently, we have

- $SAT(\neg E(\text{true } U\ p_{1,1})) = \{s_5\}$,
- $SAT(p_{1,2} \wedge \neg E(\text{true } U\ p_{1,1})) = \{s_1, s_5\} \cap \{s_5\} = \{s_5\}$,
- $SAT(\neg(p_{1,2} \wedge \neg E(\text{true } U\ p_{1,1}))) = \{s_0, s_1, s_2, s_3, s_4\}$.

Finally, according to Algorithm 5.3 we have

- $SAT(EG(\neg(p_{1,2} \wedge \neg E(\text{true } U\ p_{1,1})))) = \{s_0, s_1, s_2, s_3, s_4\}$.

Therefore, this formula is valid in this Petri net since s_0 is a satisfiable state.

When a Petri net is unbounded and thus its reachability graph is infinite, the above algorithms are infeasible in applications. Here, they are only suitable for bounded Petri nets.

The complexity of these algorithms is obviously polynomial-time with respect to the size of a reachability graph and the length of a formula [1, 2, 5]. However, the state space is often increased exponentially with the increasing of the size of a system, bringing a big challenge for the application of model checking.

Next, we introduce such a technique dealing with the state space explosion problem and apply it into the Petri-net-based model checking of CTL.

1.	$X \leftarrow SAT(\text{true}) = \{s_0, s_1, s_2, s_3, s_4, s_5\}$	$Y \leftarrow SAT(p_{1,1}) = \{s_0, s_3, s_4\}$	$Z \leftarrow \{s_0, s_2, s_3, s_4\}$
2.	$X \leftarrow X \backslash Z = \{s_1, s_5\}$	$Y \leftarrow Y \cup Z = \{s_0, s_2, s_3, s_4\}$	$Z \leftarrow \{s_1\}$
3.	$X \leftarrow X \backslash Z = \{s_5\}$	$Y \leftarrow Y \cup Z = \{s_0, s_1, s_2, s_3, s_4\}$	$Z \leftarrow \varnothing$

Fig. 5.3 The execution process of Algorithm 5.4 for the formula **E(true U** $p_{1,1}$**)**. The first line shows the initialisation, and the loop body of **while** is executed twice as shown in the second and third lines

5.2   Reduced Ordered Binary Decision Diagrams

This section first reviews the definition of reduced ordered binary decision diagram (ROBDD), introduces how to use it to encode all truth assignments of a Boolean function, and then presents how to construct an ROBDD for a given Boolean function. Finally, how to handle some manipulations of Boolean functions through their ROBDDs is introduced briefly. Readers can find more details in [6–8].

5.2.1   Reduced Ordered Binary Decision Diagrams for Encoding Boolean Functions

Here we recall the concepts related to ROBDD and introduce how to represent a Boolean function through an ROBDD.

5.2.1.1   ROBDD

Definition 5.4 (*Binary Decision Diagram*) A *binary decision diagram* is a rooted directed acyclic graph $(V, E, X, var, low, high)$ where

1. $V = \{v_0, v_1, \ldots, v_m, v^0, v^1\}$ is the set of nodes in which there is only one root node v_0 and exactly two terminal nodes v^0 and v^1,[1]
2. $E \subseteq (V \setminus \{v^0, v^1\} \times V)$ is the set of directed edges such that each non-terminal node has exactly two outgoing edges,
3. $X = \{x_1, \ldots, x_n\}$ is the set of Boolean variables,
4. $var: V \rightarrow X \cup \{0, 1\}$ is a label function such that

$$var(v^0) = 0, \ var(v^1) = 1, \ \text{and} \ var(v) \in X, \forall v \in V \setminus \{v^0, v^1\},$$

5. $low: V \setminus \{v^0, v^1\} \rightarrow V$ is the low-successor function, and
6. $high: V \setminus \{v^0, v^1\} \rightarrow V$ is the high-successor function.

The two terminal nodes are labelled by 1 and 0, respectively, representing Boolean constants **true** and **false**. Each non-terminal node is labelled with a Boolean variable and has exactly two outgoing edges associated with its low-successor and high-successor, respectively.

---

[1] When a binary decision diagram represents the Boolean function $f = $ **ture** (respectively, $f = $ **false**), it has only one node, namely v^1 (respectively, v^0), that is also the root node. For any other Boolean function, a binary decision diagram has both v^1 and v^0.

For a non-terminal node labelled by a Boolean variable, its outgoing edge associated with its low-successor (respectively, high-successor) represents that the variable is assigned **false** (respectively, **true**).

Definition 5.5 (*Ordered Binary Decision Diagram*) A binary decision diagram $(V, E, X, var, low, high)$ is called an *ordered binary decision diagram* if there is a total order of Boolean variables, say

$$x_1 < x_2 < \cdots < x_n < 0 < 1,$$

such that for each non-terminal node v we have

$$var(v) < var(low(v)) \text{ and } var(v) < var(high(v)).$$

An ordered binary decision diagram (OBDD) means that all Boolean variables of the binary decision diagram are totally ordered and each path from the root node to a terminal node visits these variables in ascending order.

Definition 5.6 (*Reduced Ordered Binary Decision Diagram*) An ordered binary decision diagram $(V, E, X, var, low, high)$ is called a *reduced ordered binary decision diagram* (ROBDD) if the following two conditions hold:

1. no two distinct nodes v_1 and v_2 are labelled by the same variable name and have the same low-successor and high-successor, i.e.,

$$\Big(var(v_1) = var(v_2) \wedge low(v_1) = low(v_2) \wedge high(v_1) = high(v_2)\Big) \Rightarrow v_1 = v_2,$$

2. no non-terminal node v has identical low-successor and high-successor, i.e.,

$$low(v) \neq high(v).$$

The two conditions in Definition 5.6 mean the uniqueness and non-redundant-ness, respectively. They guarantee that any two distinct nodes stand for two different Boolean sub-functions that will be seen later.

Example 5.8 Figure 5.4a shows an OBDD but it is not an ROBDD since it has a redundant node, namely v_5. The Boolean expression represented by the node v_5 is

$$(x_3 \vee \neg x_3) \wedge x_4 = x_4,$$

while the expression represented by the node v_4 is also x_4. Figure 5.4b is an ROBDD by deleting the redundant node v_5 out of (a) where the ordering is $x_1 < x_2 < x_3 < x_4$. Figure 5.4c is also an ROBDD where the ordering is $x_3 < x_4 < x_1 < x_2$.

The outgoing edge connecting a node and its low-successor (respectively, high-successor) is shown as a dotted (respectively, solid) line in a picture, e.g. $low(v_0) = v_1$ and $high(v_0) = v_2$ in Fig. 5.4a. Because the nodes in these pictures are listed from the top to the bottom according to the variable ordering, we omit the arrow of each directed edge.

> An (R)OBDD corresponds to a Boolean function.

Example 5.9 All the (R)OBDDs in Figs. 5.4 represent the following Boolean function:

$$
\begin{aligned}
- \; f(x_1, x_2, x_3, x_4) \; = \; & (\neg x_1 \;\wedge\; x_2 \;\wedge\; \neg x_3 \;\wedge\; x_4) \\
\vee \; & (\; x_1 \;\wedge\; \neg x_2 \;\wedge\; \neg x_3 \;\wedge\; x_4) \\
\vee \; & (\; x_1 \;\wedge\; x_2 \;\wedge\; \neg x_3 \;\wedge\; x_4) \\
\vee \; & (\; x_1 \;\wedge\; x_2 \;\wedge\; x_3 \;\wedge\; x_4) \qquad\qquad \text{Fig. 5.4 a}
\end{aligned}
$$

$$
\begin{aligned}
= \; & (\neg x_1 \;\wedge\; x_2 \;\wedge\; \neg x_3 \;\wedge\; x_4) \\
\vee \; & (\; x_1 \;\wedge\; \neg x_2 \;\wedge\; \neg x_3 \;\wedge\; x_4) \\
\vee \; & (\; x_1 \;\wedge\; x_2 \;\wedge\; x_4) \qquad\qquad\qquad\;\; \text{Fig. 5.4 b}
\end{aligned}
$$

$$
\begin{aligned}
= \; & (\; x_3 \;\wedge\; x_4 \;\wedge\; x_1 \;\wedge\; x_2) \\
\vee \; & (\neg x_3 \;\wedge\; x_4 \;\wedge\; \neg x_1 \;\wedge\; x_2) \\
\vee \; & (\neg x_3 \;\wedge\; x_4 \;\wedge\; x_1) \qquad\qquad\qquad\; \text{Fig. 5.4c}
\end{aligned}
$$

Fig. 5.4 Three OBDDs of Boolean function f in Example 5.9: **a** an OBDD but not an ROBDD, using the ordering $x_1 < x_2 < x_3 < x_4$, **b** an ROBDD using the same ordering with **a**, and **c** an ROBDD using the ordering $x_3 < x_4 < x_1 < x_2$

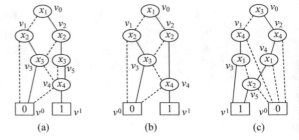

Table 5.1 Truth table of $(\neg x_1 \wedge x_2 \wedge \neg x_3 \wedge x_4) \vee (x_1 \wedge \neg x_2 \wedge \neg x_3 \wedge x_4) \vee (x_1 \wedge x_2 \wedge x_4)$

$f(x_1, x_2, x_3, x_4)$	x_1	x_2	x_3	x_4
1	0	1	0	1
1	1	0	0	1
1	1	1	0	1
1	1	1	1	1
0	Otherwise			

whose truth table is shown in Table 5.1 for verifying the above equivalence conveniently.

Each path from the root node to a terminal node in an (R)OBBD corresponds to a group of truth assignments to variables, and these assignments make the value of the function equal to **true** (if the terminal node of a path is labelled by 1) or **false** (if the terminal node of a path is labelled by 0), and all such paths exactly represent all the assignment cases as well as the corresponding values of the function.

Example 5.10 In the ROBDD in Fig. 5.4b, the path

$$v_0 \dashrightarrow v_1 \longrightarrow v_3 \dashrightarrow v_4 \longrightarrow v^1$$

represents such a truth assignment that $x_1 = x_3 = 0$ and $x_2 = x_4 = 1$, making $f = 1$. Another path

$$v_0 \dashrightarrow v_1 \dashrightarrow v^0$$

represents a group of truth assignments in which $x_1 = x_2 = 0$ and others may be any one of 0 and 1, and all of them lead to $f = 0$.

5.2.1.2 Generating a Boolean Function via an ROBDD

How to generate the related Boolean function for a given ROBDD? It is easy to achieve this goal via the following recursive formula:

$$f^v = \begin{cases} \left(var(v) \wedge f^{high(v)} \right) \vee \left(\neg var(v) \wedge f^{low(v)} \right) & v \in V \setminus \{v^0, v^1\} \\ \textbf{true} & v = v^1 \\ \textbf{false} & v = v^0 \end{cases}.$$

For convenience, we use f^v to denote the Boolean function corresponding to node v. In fact, such a Boolean function corresponds to the sub-ROBDD in which v is its root node. Sometimes, we use the name of the root node of an ROBDD to stand for the name of the ROBDD for convenience.

Example 5.11 For the ROBDD in Fig. 5.4b, we compute in turn its Boolean functions corresponding to all nodes as follows:

- $f^{v^0} = $ **false**,
- $f^{v^1} = $ **true**,
- $f^{v_4} = (x_4 \wedge f^{v^1}) \vee (\neg x_4 \wedge f^{v^0}) = (\neg x_4 \wedge $ **false**$) \vee (x_4 \wedge $ **true**$) = x_4$,
- $f^{v_3} = (x_3 \wedge f^{v^0}) \vee (\neg x_3 \wedge f^{v_4}) = (x_3 \wedge $ **false**$) \vee (\neg x_3 \wedge x_4) = \neg x_3 \wedge x_4$,
- $f^{v_2} = (x_2 \wedge f^{v_4}) \vee (\neg x_2 \wedge f^{v_3}) = (x_2 \wedge x_4) \vee (\neg x_2 \wedge \neg x_3 \wedge x_4)$,
- $f^{v_1} = (x_2 \wedge f^{v_3}) \vee (\neg x_2 \wedge f^{v^0}) = (x_2 \wedge \neg x_3 \wedge x_4) \vee (\neg x_2 \wedge $ **false**$) = x_2 \wedge \neg x_3 \wedge x_4$,
- $f^{v_0} = (x_1 \wedge f^{v_2}) \vee (\neg x_1 \wedge f^{v_1})$

$$= (x_1 \wedge ((x_2 \wedge x_4) \vee (\neg x_2 \wedge \neg x_3 \wedge x_4))) \vee (\neg x_1 \wedge x_2 \wedge \neg x_3 \wedge x_4)$$

$$= (x_1 \wedge x_2 \wedge x_4) \vee (x_1 \wedge \neg x_2 \wedge \neg x_3 \wedge x_4) \vee (\neg x_1 \wedge x_2 \wedge \neg x_3 \wedge x_4).$$

5.2.2 Constructing the ROBDD of a Boolean Function

What we pay more attention to is how to generate an ROBDD for a given Boolean function since we hope to use such a data structure as a compact representation of all possible assignments of the Boolean function. Later we will show that for a Boolean function, there is exactly one ROBDD representing it when we provide a fixed ordering of Boolean variables of this Boolean function.

5.2.2.1 If-Then-Else Operator and Shannon Expansion

Before introducing such a method, we first need to recall the *if-then-else* operator in Boolean expressions [8] as follows:

$$x \to f_1, f_2 \triangleq (x \wedge f_1) \vee (\neg x \wedge f_2)$$

In an expression of *if-then-else*, x is a Boolean variable and f_1 and f_2 are two Boolean expressions.

Property 5.1 ([8]) $x \to f_1, f_2$ *is true if and only if x and f_1 are both true or x is false and f_2 is true.*

It has been proven that any Boolean expression only including operators $\neg$, $\vee$ and $\wedge$ can be transferred into an equivalent form that only includes the *if-then-else* operator [8], due to the following equivalences:

$$- \; x = x \to 1, 0$$
$$- \; \neg x = x \to 0, 1$$
$$- \; x \vee y = x \to 1, (y \to 1, 0)$$
$$- \; x \wedge y = x \to (y \to 1, 0), 0$$

In a transfer process, the well-known *Shannon expansion* plays a critical role. Let f be a Boolean function and x be a variable in f. We denote $f[x \rightsquigarrow 1]$ (respectively, $f[x \rightsquigarrow 0]$) as the resulting function by replacing x with 1 (respectively, 0) in f. Then,

$$f = x \to f[x \rightsquigarrow 1], f[x \rightsquigarrow 0].$$

Through the Shannon expansion, a Boolean function can be transferred into a group of *if-else-then* forms.

Example 5.12   We consider the Boolean function

$$
\begin{aligned}
- \; f(x_1, x_2, x_3, x_4) \quad = \quad & (\; \neg x_1 \;\wedge\; x_2 \;\wedge\; \neg x_3 \;\wedge\; x_4 \;) \\
& \vee (\; x_1 \;\wedge\; \neg x_2 \;\wedge\; \neg x_3 \;\wedge\; x_4 \;) \\
& \vee (\; x_1 \;\wedge\; x_2 \;\wedge\; \neg x_3 \;\wedge\; x_4 \;) \\
& \vee (\; x_1 \;\wedge\; x_2 \;\wedge\; x_3 \;\wedge\; x_4 \;),
\end{aligned}
$$

and use the Shannon expansions to transfer it into an *if-then-else* form under considering x_1, x_2, x_3 and x_4 in turn. We finally obtain the following expressions:

$$
\begin{aligned}
- \; f &= x_1 \to f_1, f_2 \\
- \; f_1 &= x_2 \to f_3, f_4 \\
- \; f_2 &= x_2 \to f_4, 0 \\
- \; f_3 &= x_3 \to f_5, f_5 \\
- \; f_4 &= x_3 \to 0, f_5 \\
- \; f_5 &= x_4 \to 1, 0
\end{aligned}
$$

In this transfer process, the computations and uses of f_1, f_2, f_3, f_4 and f_5 are listed as follows:

$$- \; f_1 = f[x_1 \rightsquigarrow 1] = (\neg x_2 \wedge \neg x_3 \wedge x_4) \vee (x_2 \wedge \neg x_3 \wedge x_4) \vee (x_2 \wedge x_3 \wedge x_4),$$
$$- \; f_2 = f[x_1 \rightsquigarrow 0] = x_2 \wedge \neg x_3 \wedge x_4,$$
$$- \; f_3 = f_1[x_2 \rightsquigarrow 1] = (\neg x_3 \wedge x_4) \vee (x_3 \wedge x_4),$$
$$- \; f_4 = f_1[x_2 \rightsquigarrow 0] = f_2[x_2 \rightsquigarrow 1] = \neg x_3 \wedge x_4,$$

Algorithm 5.5: OBDD(f, k): A Recursive Algorithm of Generating an OBDD of a Boolean Function.

Input: A Boolean function $f(x_k, x_{k+1}, \cdots, x_n)$ and k; Notice, we assume that the terminal nodes v^0 and v^1 have been created before calling **OBDD**($f, 1$) and the Boolean variable ordering is $x_1 < x_2 < \cdots < x_n$.

Output: The root node v_k;

1   **if** $f = $ **true then**
2 | **return** v^1;

3   **else**
4 | **if** $f = $ **false then**
5 | | **return** v^0;

6 | **else**
7 | | Create a node v_k;
8 | | $var(v_k) \leftarrow x_k$;
9 | | $high(v_k) \leftarrow$ **OBDD**($f[x_k \rightsquigarrow 1], k + 1$);
10 | | $low(v_k) \leftarrow$ **OBDD**($f[x_k \rightsquigarrow 0], k + 1$);
11 | | **return** v_k;

- $f_5 = f_3[x_3 \rightsquigarrow 1] = f_3[x_3 \rightsquigarrow 0] = f_4[x_3 \rightsquigarrow 0] = x_4$.

Obviously, expressions f, f_1, f_2, f_3, f_4 and f_5 exactly correspond to nodes v_0, v_2, v_1, v_5, v_3 and v_4 in Fig. 5.4a, respectively. In other words, the OBDD in Fig. 5.4a is easily constructed by Algorithm 5.5 according to the above expressions since there is an explicit correspondence between them. However, we have known that the node v_5 is redundant in this OBDD, and now we know it is caused just by the following Boolean function

- $f_3 = f_1[x_2 \rightsquigarrow 1] = (\neg x_3 \wedge x_4) \vee (x_3 \wedge x_4)$

which means that f_3 and f_5 have the same value for any assignment to x_3.

Therefore, when a Boolean function written in the *if-then-else* form has such a sub-expression:

$$f' = x \rightarrow f'', f'',$$

i.e., the *then* part and the *else* part are identical, we can replace all f' with f'' and delete f'. Similarly, if there are two sub-functions whose expressions are identical, then we only retain one of them. Through these operations, we can obtain a reduced expression for any Boolean function.

Example 5.13 For the above example, we have:

- $f = x_1 \rightarrow f_1, f_2$
- $f_1 = x_2 \rightarrow f_5, f_4$
- $f_2 = x_2 \rightarrow f_4, 0$
- $f_4 = x_3 \rightarrow 0, f_5$
- $f_5 = x_4 \rightarrow 1, 0$

and if we use Algorithm 5.5 to construct an OBDD based on this expression, then we obtain the ROBDD in Fig. 5.4b.

5.2.2.2 Uniqueness of ROBDD and Generation Algorithm of ROBDD

It can be imagined that based on the reduced expressions written in the *if-then-else* form, we can use Algorithm 5.5 to construct an ROBDD for any Boolean function. Additionally, when we use a given variable ordering to generate the *if-then-else* expression of a Boolean function, the reduced expression is obviously unique, which means that there is a unique ROBDD representing it.

Theorem 5.1 ([8]) *Given a variable ordering* $x_1 < x_2 < \cdots < x_n$ *and a Boolean function* $f(x_1, x_2, \ldots, x_n)$, *there is exactly one ROBDD* v *such that*

$$f^v = f(x_1, x_2, \ldots, x_n).$$

Proof It can be proven by induction on the number of arguments of f. It is easy to check that when $n = 0$, there only two Boolean functions $f = $ **true** and $f = $ **false**. The former (respectively, the latter) corresponds to the ROBDD in which there is only one node, namely v^1 (respectively, v^0), that is also the root node. When $n = 1$, there are four Boolean functions as follows:

- $f(x) = $ **true**. For this case, the ROBDD includes only one node that is both labelled by x as the root and labelled by 1.
- $f(x) = $ **false**. For this case, the ROBDD also includes only one node that is both labelled by x as the root and labelled by 0.
- $f(x) = x$. For this case, the ROBDD has three nodes: the root labelled by x, the terminal node v^0 as the low-successor of the root, and the terminal node v^1 as the high-successor of the root.
- $f(x) = \neg x$. For this case, the ROBDD also has three nodes: the root labelled by x, the terminal node v^0 as the high-successor of the root, and the terminal node v^1 as the low-successor of the root.

We assume that for any Boolean function with n arguments this conclusion holds, and we now consider the case of $n + 1$ arguments. We denote

$$f_b(x_2, \ldots, x_{n+1}) \triangleq f(b, x_2, \ldots, x_{n+1}), \ b \in \mathbb{N}_1 = \{0, 1\}$$

Algorithm 5.6: ROBDD(f, k): A Recursive Algorithm of Generating the ROBDD of a Boolean Function.

Input: A Boolean function $f(x_k, x_{k+1}, \cdots, x_n)$ and k; Notice, we assume that the terminal nodes v^0 and v^1 have been created before calling **ROBDD**$(f, 1)$ and the Boolean variable ordering is $x_1 < x_2 < \cdots < x_n$.

Output: The root node v_k;

1 **if** $f = $ **true then**
2 | **return** v^1;

3 **else**
4 | **if** $f = $ **false then**
5 | | **return** v^0;

6 | **else**
7 | | Create a node v_k;
8 | | $var(v_k) \leftarrow x_k$;
9 | | $high(v_k) \leftarrow$ **ROBDD**$(f[x_k \rightsquigarrow 1], k+1)$;
10 | | $low(v_k) \leftarrow$ **ROBDD**$(f[x_k \rightsquigarrow 0], k+1)$;
11 | | **if** $low(v_k) = high(v_k)$ **then**
12 | | | $v \leftarrow low(v_k)$;
13 | | | Delete v_k;
14 | | | **return** v;

15 | | **else**
16 | | | **if** *there is a node v in the existing sub-diagram such that*

$$var(v) = var(v_k) \wedge high(v) = high(v_k) \wedge low(v) = low(v_k)$$

            then
17 | | | | Delete v_k;
18 | | | | **return** v;

19 | | | **else**
20 | | | | **return** v_k;

---

and use the ordering $x_1 < x_2 < \cdots < x_{n+1}$. We also assume that the *if-then-else* expressions of them are reduced. According to the Shannon expansion we have

$$f(x_1, x_2, \ldots, x_{n+1}) = x_1 \rightarrow f_1(x_2, \ldots, x_{n+1}), f_0(x_2, \ldots, x_{n+1}).$$

Since $f_1 \neq f_0$, we can construct two different nodes v_0 and v_1 corresponding to f_0 and f_1, respectively. And v_0 and v_1 are two ROBDDs that both have a uniqueness according to the assumption. Therefore, we construct a node v such that it is labelled by x_1 and its low- and high-successors are v_0 and v_1, respectively. Obviously, v represents f and is of the uniqueness.   □

In [8], an effective algorithm is presented to generate the ROBBD of a Boolean function. It is similar to Algorithm 5.5, but it goes to search and decide if a sub-function is redundant or identical to an existing one during such a recursive procedure. Here we present a similar recursive procedure (Algorithm 5.6). Although it is not as effective as the one in [8], it can be understood more easily. Its idea is described as follows:

For a node, its high-successor and low-successor are recursively constructed in turn and then this node as a high-successor or a low-successor will be returned to the calling level, unless one of the following two cases happens:
– Its high-successor and low-successor are the same node. For this case, this node (i.e., this high-/low-successor) as the returned value is transmitted to the calling level and it is deleted, as shown in Lines 11–14 in Algorithm 5.6.
– An existing node is equal to this node. For this case, this existing one as the returned value is transmitted to the calling level and this node is deleted, as shown in Lines 16–18 in Algorithm 5.6.

Example 5.14 We continue to consider the construction of the ROBDD of the function in Example 5.12 according to Algorithm 5.6. Figure 5.5 shows the recursive process of constructing this ROBDD, including multiple steps of deleting some generated nodes. As shown in Fig. 5.5e, when the left node labeled by x_4 is created as the low-successor of the node labeled by x_3, the algorithm checks that there has been a node, namely the right node labeled by x_4 in Fig. 5.5e, is equal to it; and thus the left one is deleted and the right one is as the low-successor of the node labeled by x_3, as shown in Fig. 5.5f. Continually, before the node labeled by x_3 in Fig. 5.5f is transmitted to the calling level (i.e., it is created as the high-successor of the node labeled by x_2 in Fig. 5.5 g), the algorithm checks that its high-successor and low-successor are equal, and thus its high-/low-successor is transmitted to the calling level (i.e. the node labeled by x_4 is as the high-successor of the node labeled by x_2 in Fig. 5.5g) and it is deleted. Similarly, readers can understand the creations or deletions of other nodes in Fig. 5.5.

Question

Try to analyse the complexity of Algorithms 5.5 and 5.6. Try to write a more efficient algorithm.

Fig. 5.5 The recursive construction procedure of the ROBDD of the function in Example 5.12

5.2.3 Operations of Boolean Functions by Manipulating ROBDDs

How to implement the operations of Boolean functions through their ROBDDs? We concern the following ones:

1. the negation of a Boolean function,
2. changing the sign of a variable of a Boolean function, and
3. computing the conjunction or disjunction of two Boolean functions.

5.2.3.1 Negation of Boolean Function

Given a Boolean function f and its negation $\neg f$, we know that a truth assignment makes $f = \textbf{true}$ if and only if it makes $\neg f = \textbf{false}$. Therefore, it is very easy to implement the negation of a Boolean function:

> Given the ROBDD of a Boolean function, exchanging the labels of the two terminal nodes leads to the ROBDD of the negation of this Boolean function.

5.2.3.2 Changing the Sign of a Variable of a Boolean Function

Given a Boolean function $f(x_1, x_2, \ldots, x_n)$ and its ROBDD, if we change the sign of some variable (say x_k) of the expression of f, then we obtain another Boolean function f' where

$$f'(x_1, \ldots, x_k, \ldots, x_n) = f(x_1, \ldots, \neg x_k, \ldots, x_n).$$

What is the ROBDD of f'? Because of the following facts:

- $f' = x_k \rightarrow f'[x_k \rightsquigarrow 1], f'[x_k \rightsquigarrow 0]$,
- $f'[x_k \rightsquigarrow 1] = f[x_k \rightsquigarrow 0]$,
- $f'[x_k \rightsquigarrow 0] = f[x_k \rightsquigarrow 1]$,

we have

$$f' = x_k \rightarrow f[x_k \rightsquigarrow 0], f[x_k \rightsquigarrow 1].$$

> Therefore, exchanging the low-successors and high-successors of all the nodes labelled by x_k in the ROBDD of f leads to the ROBDD of f'.

For convenience, we use the notation

$$f[x \rightsquigarrow \neg x]$$

to represent the Boolean function generated by changing the sign of x of f.

For the above two manipulations, readers can construct some instances to test them, and here we do not provide some related examples.

Algorithm 5.7: AND(v, v'): A Recursive Algorithm of Generating the ROBDD of the Conjunction of Two Boolean Functions via Their ROBDDs.

Input: Two nodes v and v' from two ROBDDs; Notice, we assume that the terminal nodes v^0 and v^1 have been created before calling **AND**(v_0, v'_0) where v_0 and v'_0 represent the root nodes of the ROBDDs of the given two Boolean functions.

Output: The root node $v : v'$;

1   **if** $var(v) = \textbf{\textit{true}} \wedge var(v') = \textbf{true}$ **then**
2 | **return** v^1;

3   **else**
4 | **if** $(var(v) = \textbf{true} \vee var(v) = \textbf{false}) \wedge (var(v') = \textbf{true} \vee var(v') = \textbf{false})$ **then**
5 | | **return** v^0;

6 | **else**
7 | | Create a node $v : v'$;
8 | | **if** $var(v) = var(v')$ **then**
9 | | | $var(v : v') \leftarrow var(v)$;
10 | | | $high(v : v') \leftarrow$ **AND**($high(v)$, $high(v')$);
11 | | | $low(v : v') \leftarrow$ **AND**($low(v)$, $low(v')$);

12 | | **else**
13 | | | **if** $var(v) < var(v')$ **then**
14 | | | | $var(v : v') \leftarrow var(v)$;
15 | | | | $high(v : v') \leftarrow$ **AND**($high(v)$, v');
16 | | | | $low(v : v') \leftarrow$ **AND**($low(v)$, v');

17 | | | **else**
18 | | | | $var(v : v') \leftarrow var(v')$;
19 | | | | $high(v : v') \leftarrow$ **AND**(v, $high(v')$);
20 | | | | $low(v : v') \leftarrow$ **AND**(v, $low(v')$);

21 | | **if** $low(v : v') = high(v : v')$ **then**
22 | | | $v'' \leftarrow low(v : v')$;
23 | | | Delete $v : v'$;
24 | | | **return** v'';

25 | | **else**
26 | | | **if** *there is a node v'' in the existing sub-diagram such that*

$$var(v'') = var(v : v') \wedge high(v'') = high(v : v') \wedge low(v'') = low(v : v')$$

              then
27 | | | | Delete $v : v'$;
28 | | | | **return** v'';

29 | | | **else**
30 | | | | **return** $v : v'$;

5.2.3.3   Conjunction of Two Boolean Functions

In what follows, we introduce how to implement the conjunction of two Boolean functions over their ROBDDs. The disjunction and other binary operators are all similar and omitted here.

We first assume that the ROBDDs of two Boolean functions are created according to the same variable ordering.

Given two Boolean functions $f(x_1, x_2, \ldots, x_n)$ and $g(x_1, x_2, \ldots, x_n)$, their conjunction is denoted as

$$h(x_1, x_2, \ldots, x_n) = f(x_1, x_2, \ldots, x_n) \wedge g(x_1, x_2, \ldots, x_n).$$

According to the Shannon expansion, we have

$$- h = x \rightarrow h[x \rightsquigarrow 1], h[x \rightsquigarrow 0]$$
$$= x \rightarrow (f[x \rightsquigarrow 1] \wedge g[x \rightsquigarrow 1]), (f[x \rightsquigarrow 0] \wedge g[x \rightsquigarrow 0]).$$

The above formula illustrates a recursive relation obviously. Hence, we can start at the roots of the ROBDDs of f and g to recursively construct the ROBDD of h as shown in Algorithm 5.7. When scanning two nodes of the ROBDDs of f and g, we first create a new node, and then consider the following two cases to recursively generate its high-successor and low-successor:

– If the two scanned nodes have the same label, the new node is labelled by this label and then recursively generate its high-successor (respectively, low-successor) through simultaneously scanning their high-successors (respectively, low-successors). It is shown in Lines 8–11 in Algorithm 5.7.
– If the two scanned nodes have different labels, the new node's label is assigned the less ordering value and then recursively generate its high-successor (respectively low-successor) through scanning the high-successor (respectively, low-successor) of the node with the less ordering value while keeping the node with the greater ordering value unchanged. It is shown in Lines 13–20 in Algorithm 5.7. Notice: the conduction of this case is correct because for the scanned node with the greater ordering value, those variables whose ordering values are less than this node's can be assigned an arbitrary truth but not change the Boolean function's value in the related ROBDD.

Similar to Algorithm 5.6, before returning a created node to the calling level, we check if it should be reduced, as shown in Lines 21–30 in Algorithm 5.7. Given two nodes v and v' coming from two ROBDDs, the constructed node corresponding to them is named as $v : v'$ for simplicity.

Example 5.15 We consider the construction of the ROBDD of the conjunction of the following two Boolean functions:

$$f(x_1, x_2, x_3, x_4) = (\neg x_1 \wedge x_2 \wedge \neg x_3 \wedge x_4) \vee (x_1 \wedge \neg x_2 \wedge \neg x_3 \wedge x_4) \vee (x_1 \wedge x_2 \wedge x_4)$$

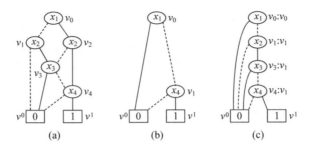

(a)   (b)   (c)

Fig. 5.6 a The ROBDD of the Boolean function f in Example 5.15, **b** the ROBDD of the Boolean function g, and **c** the ROBDD of $f \wedge g$. Note, the variable ordering is $x_1 < x_2 < x_3 < x_4$

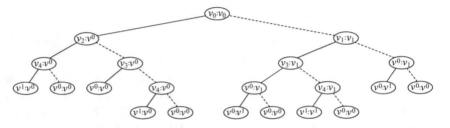

Fig. 5.7 The process of recursively scanning the two ROBDDs in Fig. 5.6a, b according to Algorithm 5.7

and

$$g(x_1, x_2, x_3, x_4) = (\neg x_1 \wedge x_4).$$

The ROBDD of f has been illustrated in Fig. 5.4b, but for readability, we show it in Fig. 5.6a again. The ROBDD of g is shown in Fig. 5.6b. The ROBDD of their conjunction is shown in Fig. 5.6c. Figure 5.7 unfolds the recursively process according to Algorithm 5.7, while Fig. 5.8 clearly illustrates the process of generating and reducing nodes.

5.2.3.4   UCDD Package and Optimal Ordering

In fact, there are a number of successful tools of generating ROBDDs of Boolean functions as well as implementing all kinds of Boolean operations over ROBDDs, such as the tool package UCDD [9]. UCDD (Colorado University Decision Diagram) defines a data structure type BDD to specify and store a Boolean function and presents all kinds of Boolean operators like ! ($\neg$), + ($\vee$) and * ($\wedge$) that can be operated on one or two variables of this data type. These contents are beyond the scope of this book and thus not be introduced any more.

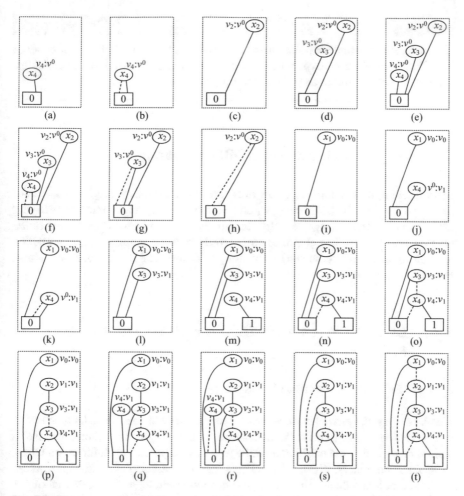

Fig. 5.8 The process of generating and reducing nodes for the construction of the ROBDD in Fig. 5.6c based on scanning the ROBDDs in Fig. 5.6a, b

The following two facts need to be clarified before the end of this section:

- For a given Boolean function, two different orderings of variables lead to two different ROBDDs of different sizes as shown in Fig. 5.4b, c.
- In theory, an improper ordering can result in the node explosion problem [10], i.e., the number of nodes in an ROBDD grows exponentially with the number of variables [11], whereas verifying if an ordering is optimal is NP-complete [12].

Try to construct a Boolean function and its variable ordering such that the number of its ROBDD nodes grows exponentially with the increase of the number of variables. Why is the verification of an optimal ordering so hard?

5.3   Verifying CTL with the ROBDD Technique

In this section, we first introduce how to encode the reachability graph of a safe Petri net based on the ROBDD technique, and then present the algorithms of verifying CTL formulae. Especially, we only need to encode all reachable markings rather than a whole reachability graph. In other words, we do not encode those transition relation among markings so that lots of storage space can be saved. For more details, readers can refer to [11–14].

5.3.1   *Encoding Reachability Graphs of Safe Petri Nets Using ROBDDs*

5.3.1.1   Encoding All Reachable Markings

If a place of a safe Petri net is viewed as a Boolean variable, then a reachable marking can be viewed as a conjunction term of these variables or their negations such that if a place is marked in the marking then the related variable occurs in the conjunction, or its negation occurs. Then, all reachable markings are represented as a Boolean function that is the disjunction of these conjunction terms, and a (safe) marking is reachable if and only if the truth assignment corresponding to the marking makes the Boolean function true. Here, the truth assignment corresponding to a safe marking means that if a place is marked at the marking then the corresponding variable is assigned 1, or it is assigned 0.

How to encode the reachable markings of a bounded but unsafe Petri net?

Example 5.16 We consider the reachability graph in Fig. 5.1a and can use the following Boolean function written in a disjunction normal form to represent the 6 markings:

$- f(p_{1,1}, p_{1,2}, p_{1,3}, r_1, r_2, p_{2,1}, p_{2,2}, p_{2,3}) =$

$$
\begin{aligned}
(& \; p_{1,1} \wedge \neg p_{1,2} \wedge \neg p_{1,3} \wedge \; r_1 \wedge \; r_2 \wedge \; p_{2,1} \wedge \neg p_{2,2} \wedge \neg p_{2,3}) \\
\vee (& \; p_{1,1} \wedge \neg p_{1,2} \wedge \neg p_{1,3} \wedge \; r_1 \wedge \neg r_2 \wedge \neg p_{2,1} \wedge \; p_{2,2} \wedge \neg p_{2,3}) \\
\vee (& \; p_{1,1} \wedge \neg p_{1,2} \wedge \neg p_{1,3} \wedge \neg r_1 \wedge \neg r_2 \wedge \neg p_{2,1} \wedge \neg p_{2,2} \wedge \; p_{2,3}) \\
\vee (\neg & p_{1,1} \wedge \; p_{1,2} \wedge \neg p_{1,3} \wedge \neg r_1 \wedge \; r_2 \wedge \; p_{2,1} \wedge \neg p_{2,2} \wedge \neg p_{2,3}) \\
\vee (\neg & p_{1,1} \wedge \neg p_{1,2} \wedge \; p_{1,3} \wedge \neg r_1 \wedge \neg r_2 \wedge \; p_{2,1} \wedge \neg p_{2,2} \wedge \neg p_{2,3}) \\
\vee (\neg & p_{1,1} \wedge \; p_{1,2} \wedge \neg p_{1,3} \wedge \neg r_1 \wedge \neg r_2 \wedge \neg p_{2,1} \wedge \; p_{2,2} \wedge \neg p_{2,3}).
\end{aligned}
$$

We use the ordering

$$p_{1,1} < p_{1,2} < p_{1,3} < r_1 < r_2 < p_{2,1} < p_{2,2} < p_{2,3}$$

and then obtain the ROBDD of f shown in Fig. 5.9. For instance, the path

$$p_{1,1} \dashrightarrow p_{1,2} \dashrightarrow p_{1,3} \longrightarrow r_1 \dashrightarrow r_2 \dashrightarrow p_{2,1} \longrightarrow p_{2,2} \dashrightarrow p_{2,3} \dashrightarrow 1$$

represents a truth assignment of variables in which variables $p_{1,3}$ and $p_{2,1}$ both are assigned 1 and others are all assigned 0, making $f = 1$ and representing the reachable marking $\{p_{1,3}, p_{2,1}\}$. For another instance, the path

$$p_{1,1} \longrightarrow p_{1,2} \dashrightarrow p_{1,3} \longrightarrow 0$$

represents a group of truth assignments in which

$$p_{1,1} = p_{1,3} = 1 \wedge p_{1,2} = 0$$

and all the others can be assigned an arbitrary value (of 0 and 1). Therefore, this path stands for $2^5 = 32$ assignment cases but all of them results in $f = 0$, which means that they are not reachable in the Petri net in Fig. 5.2.

Fig. 5.9 The ROBDD representing the reachable markings in Fig. 5.1a where the variable ordering is $p_{1,1} < p_{1,2} < p_{1,3} < r_1 < r_2 < p_{2,1} < p_{2,2} < p_{2,3}$. Note that the edges connecting non-terminal nodes and the terminal node labelled by 0 are not drawn for simplicity, and the node names are omitted too

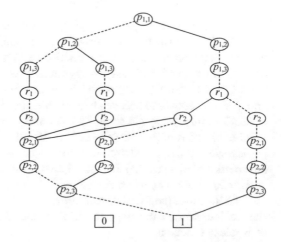

Algorithm 5.8: Generating the ROBDD Encoding All Reachable Markings of a Petri Net.

Input: A Safe Petri net (P, T, F, M_0). Note, we assume that the place ordering is
$\quad\quad p_1 < p_2 < \cdots p_n$ used in the generating process;
Output: The ROBDD encoding all reachable markings;

1 $\mathbb{M} \leftarrow$ **true**;
2 **for** $p \leftarrow p_1$ *to* p_n **do**
3 $\quad$ **if** $M_0(p) = 1$ **then**
4 $\quad\quad \lfloor \ \mathbb{M} \leftarrow \mathbb{M} \wedge p$;

5 $\quad$ **else**
6 $\quad\quad \lfloor \ \mathbb{M} \leftarrow \mathbb{M} \wedge \neg p$;

7 $New \leftarrow \mathbb{M}$;
8 **while** $New \neq$ **false do**
9 $\quad New \leftarrow$ **SUC**$(New) - \mathbb{M}$;
10 $\quad \lfloor \ \mathbb{M} \leftarrow \mathbb{M} \vee New$;

11 **return** $\mathbb{M}$;

Algorithm 5.9: SUC(f): Generating the Successors of a Given Group of Reachable Markings encoded by an ROBDD f.

Input: A group of reachable markings of the Petri net (P, T, F, M_0) encoded by an ROBDD
$\quad\quad f$;
Output: The successors of the group of markings;

1 $Suc \leftarrow$ **false**;
2 **for** *each* $t \in T$ **do**
3 $\quad Enable \leftarrow \bigwedge_{p \in {}^{\bullet}t} p \wedge f$;
4 $\quad$ **if** $Enable \neq$ **false then**
5 $\quad\quad \lfloor \ Enable \leftarrow Enable[p \rightsquigarrow \neg p]_{p \in ({}^{\bullet}t \cup t^{\bullet}) \setminus ({}^{\bullet}t \cap t^{\bullet})}$;
6 $\quad Suc \leftarrow Suc \vee Enable$;

7 **return** Suc;

From the above example, we seemly cannot see the compression advantage when using the ROBDD technique to encode all reachable markings. In fact, when a Petri net has a number of concurrent transitions and a good variable ordering is found, the compression effect is usually significant, which has been illustrated by much work [11, 13, 15]. For example of the dining 5000-philosophers problem, we take about 4500 s seconds to encode all reachable markings (about 10^{3300} markings) using a PC. Certainly, we spend about 38500 s to compute a variable ordering, while the number of nodes of this ROBDD is 219970 [11].

Algorithm 5.8 describes the generating process of the ROBDD encoding all reachable markings of a Petri net. In fact, when we use the UCDD tool package to generate such an ROBDD, we can directly define $\mathbb{M}$ as a variable of the BDD type provided by the UCDD. Since the ROBDD technique is to handle Boolean functions as shown in the previous section, the program variables specified in Algorithm 5.8 can be viewed as Boolean functions.

Lines 2–6 in Algorithm 5.8 are to encode the initial marking. *New* is a temporary variable that records those unexplored ones of all generated markings and thus is initialised by the initial marking. $\mathbf{SUC}(f)$ is to compute the successors of such markings that are represented by the Boolean function f, which is described in Algorithm 5.9. In what follows, we introduce $\mathbf{SUC}(f)$ in detail.

5.3.1.2 Batch Generation of Successors

Given a Boolean expression f representing a group of markings and a transition t, how to decide if the transition is enabled or disabled at the markings? and how to compute all successors reached by firing t?

The conjunction of all the Boolean variables corresponding to the input places of t represents the condition of enabling t, i.e., if this conjunction expression is true then t is enabled, or else it is disabled. Therefore, we take a conjunction operation between this expression and f as follows:

$$\left(\bigwedge_{p \in {}^{\bullet}t} p \right) \wedge f,$$

which exactly corresponds to Line 3 in Algorithm 5.9. If the new expression is **false**, then the transition is disabled at each one in the group of markings, or it is enabled. In fact, the new expression exactly represents those markings enabling this transition in the group.

Example 5.17 We consider the following Boolean function that represents the markings s_0, s_1 and s_2 of the reachability graph in Fig. 5.1a:

$$- f = (p_{1,1} \wedge \neg p_{1,2} \wedge \neg p_{1,3} \wedge r_1 \wedge r_2 \wedge p_{2,1} \wedge \neg p_{2,2} \wedge \neg p_{2,3}) \vee$$
$$(\neg p_{1,1} \wedge p_{1,2} \wedge \neg p_{1,3} \wedge \neg r_1 \wedge r_2 \wedge p_{2,1} \wedge \neg p_{2,2} \wedge \neg p_{2,3}) \vee$$
$$(\neg p_{1,1} \wedge \neg p_{1,2} \wedge p_{1,3} \wedge \neg r_1 \wedge \neg r_2 \wedge p_{2,1} \wedge \neg p_{2,2} \wedge \neg p_{2,3}).$$

We consider the transition $t_{2,1}$. Since ${}^{\bullet}t_{2,1} = \{r_2, p_{2,1}\}$, the markings that can both enable $t_{2,1}$ and be represented by f can be obtained by the following computation:

$$- \left(r_2 \wedge p_{2,1} \right) \wedge f = \left(r_2 \wedge p_{2,1} \right) \wedge$$
$$\Big((p_{1,1} \wedge \neg p_{1,2} \wedge \neg p_{1,3} \wedge r_1 \wedge r_2 \wedge p_{2,1} \wedge \neg p_{2,2} \wedge \neg p_{2,3}) \vee$$
$$(\neg p_{1,1} \wedge p_{1,2} \wedge \neg p_{1,3} \wedge \neg r_1 \wedge r_2 \wedge p_{2,1} \wedge \neg p_{2,2} \wedge \neg p_{2,3}) \vee$$
$$(\neg p_{1,1} \wedge \neg p_{1,2} \wedge p_{1,3} \wedge \neg r_1 \wedge \neg r_2 \wedge p_{2,1} \wedge \neg p_{2,2} \wedge \neg p_{2,3}) \Big)$$
$$= (p_{1,1} \wedge \neg p_{1,2} \wedge \neg p_{1,3} \wedge r_1 \wedge r_2 \wedge p_{2,1} \wedge \neg p_{2,2} \wedge \neg p_{2,3}) \vee$$
$$(\neg p_{1,1} \wedge p_{1,2} \wedge \neg p_{1,3} \wedge \neg r_1 \wedge r_2 \wedge p_{2,1} \wedge \neg p_{2,2} \wedge \neg p_{2,3}),$$

which exactly corresponds to s_0 and s_1. This case is coincided with the fact. When we consider another transition $t_{2,2}$ with ${}^\bullet t_{2,2} = \{r_1, p_{2,2}\}$, we have:

$$- \left(r_1 \wedge p_{2,2}\right) \wedge f = \left(r_1 \wedge p_{2,2}\right) \wedge$$
$$\Big((p_{1,1} \wedge \neg p_{1,2} \wedge \neg p_{1,3} \wedge r_1 \wedge r_2 \wedge p_{2,1} \wedge \neg p_{2,2} \wedge \neg p_{2,3}) \vee$$
$$(\neg p_{1,1} \wedge p_{1,2} \wedge \neg p_{1,3} \wedge \neg r_1 \wedge r_2 \wedge p_{2,1} \wedge \neg p_{2,2} \wedge \neg p_{2,3}) \vee$$
$$(\neg p_{1,1} \wedge \neg p_{1,2} \wedge p_{1,3} \wedge \neg r_1 \wedge \neg r_2 \wedge p_{2,1} \wedge \neg p_{2,2} \wedge \neg p_{2,3})\Big)$$
$$= \textbf{false},$$

which means that $t_{2,2}$ is disabled at any marking represented by f. This case is also coincided with the fact.

Now that the Boolean expression *Enable* in Line 3 in Algorithm 5.9 exactly represents those markings enabling a transition t, we can utilise it to compute the successors reached by firing t.

> We can easily understand the following two features from the disjunction normal form of the expression
>
> $$\left(\bigwedge_{p \in {}^\bullet t} p\right) \wedge f.$$
>
> 1. for each $p \in {}^\bullet t$, the Boolean variable p occurs in all conjunction terms of this disjunction normal form but $\neg p$ does not occur in them, since each conjunction term represents a marking enabling t;
> 2. for each $p \in t^\bullet \setminus {}^\bullet t$, the negation $\neg p$ occurs in them but p does not occur since the Petri net we consider here is safe.
>
> Hence, to compute the Boolean expression representing the successors reached by firing t is to change each p such that $p \in {}^\bullet t \setminus t^\bullet$ into $\neg p$ and each $\neg p$ such that $p \in t^\bullet \setminus {}^\bullet t$ into p. Anyway, generating these successors is to change the signs of all $p \in ({}^\bullet t \cup t^\bullet) \setminus ({}^\bullet t \cap t^\bullet)$, as shown in Line 5 in Algorithm 5.9.

Example 5.18 We continue to consider the transition $t_{2,1}$ in Example 5.17. Since ${}^\bullet t_{2,1} = \{r_2, p_{2,1}\}$ and $t_{2,1}^\bullet = \{p_{2,2}\}$, we change the signs of r_2, $p_{2,1}$ and $p_{2,2}$ in the following expression

$$- (p_{1,1} \wedge \neg p_{1,2} \wedge \neg p_{1,3} \wedge r_1 \wedge r_2 \wedge p_{2,1} \wedge \neg p_{2,2} \wedge \neg p_{2,3}) \vee$$
$$(\neg p_{1,1} \wedge p_{1,2} \wedge \neg p_{1,3} \wedge \neg r_1 \wedge r_2 \wedge p_{2,1} \wedge \neg p_{2,2} \wedge \neg p_{2,3}),$$

and then have the following expression

$$- (p_{1,1} \wedge \neg p_{1,2} \wedge \neg p_{1,3} \wedge r_1 \wedge \neg r_2 \wedge \neg p_{2,1} \wedge p_{2,2} \wedge \neg p_{2,3}) \vee$$
$$(\neg p_{1,1} \wedge p_{1,2} \wedge \neg p_{1,3} \wedge \neg r_1 \wedge \neg r_2 \wedge \neg p_{2,1} \wedge p_{2,2} \wedge \neg p_{2,3})$$

that exactly represents the successors reached by firing the transition $t_{2,1}$, i.e., s_3 and s_5 in Fig. 5.1a.

> We see another advantage of using the ROBDD technique to generate and encode all reachable markings: firing a transition can generate a group of markings at one time. As a result, it is usually much faster than a one-by-one method like Algorithm 1.1.

5.3.1.3   Generation of Newly To-be-explored Markings

In Line 9 of Algorithm 5.8, **SUC**(*New*) computes all successors of the currently-being-explored markings represented by *New*, but some of these successors maybe have been explored. After the computation of **SUC**(*New*), $\mathbb{M}$ records all explored markings. Hence, the operation

$$\mathbf{SUC}(New) - \mathbb{M}$$

is to compute those markings that have been generated but not explored. This operation means that those common markings represented both by **SUC**(*New*) and by $\mathbb{M}$ are removed out of the former. According to the set theory, it is easily understood and implemented. As a Boolean operation, however, how to implement it?

We first write **SUC**(*New*) and $\mathbb{M}$ in the disjunction normal forms as follows:

$$\mathbf{SUC}(New) = A \vee B \quad \text{and} \quad \mathbb{M} = B \vee C$$

where the marking set corresponding to B is the maximum common part of **SUC**(*New*) and $\mathbb{M}$. In other words, the marking sets corresponding to A, B and C are disjoint pairwise. Obviously, what we expect about **SUC**(*New*) $- \mathbb{M}$ is A. As we know, the expression

$$\mathbf{SUC}(New) \vee \mathbb{M} = A \vee B \vee C$$

means all markings represented by **SUC**(*New*) and $\mathbb{M}$. Consequently, we have

$$\mathbf{SUC}(New) \wedge \neg \mathbb{M} = (\mathbf{SUC}(New) \vee \mathbb{M}) \wedge \neg \mathbb{M} = (A \vee B \vee C) \wedge \neg \mathbb{M} = A \wedge \neg \mathbb{M}.$$

Because there is no any common marking between A and $\mathbb{M}$, we know that the set of all markings corresponding to A is a subset of the set of all markings corresponding to $\neg \mathbb{M}$, i.e.,

$$\mathbf{SUC}(New) \wedge \neg \mathbb{M} = A \wedge \neg \mathbb{M} = A,$$

which is exactly what we expect.

> Therefore, we have
>
> $$\mathbf{SUC}(New) - \mathbb{M} = \mathbf{SUC}(New) \wedge \neg \mathbb{M}.$$
>
> The conjunction of two Boolean functions and the negation of a Boolean function can be manipulated through their ROBDDs. Therefore, this subtract of two Boolean functions can also be handled in view of their ROBDDs.

Example 5.19 We use the Petri net in Fig. 5.2a to shown the computation process of Algorithm 5.8 as follows:

1. The initial case:

   $-\ New = \mathbb{M} = p_{1,1} \wedge \neg p_{1,2} \wedge \neg p_{1,3} \wedge r_1 \wedge r_2 \wedge p_{2,1} \wedge \neg p_{2,2} \wedge \neg p_{2,3}.$

2. The first round of the **for** loop:

   $-\ \mathbf{SUC}(New) = (\neg p_{1,1} \wedge p_{1,2} \wedge \neg p_{1,3} \wedge \neg r_1 \wedge r_2 \wedge p_{2,1} \wedge \neg p_{2,2} \wedge \neg p_{2,3}) \vee$
   $\qquad\qquad\qquad (p_{1,1} \wedge \neg p_{1,2} \wedge \neg p_{1,3} \wedge r_1 \wedge \neg r_2 \wedge \neg p_{2,1} \wedge p_{2,2} \wedge \neg p_{2,3}),$
   $-\ New = \mathbf{SUC}(New) - \mathbb{M} = \mathbf{SUC}(New),$
   $-\ \mathbb{M} = \mathbb{M} \vee New$
   $\qquad = (p_{1,1} \wedge \neg p_{1,2} \wedge \neg p_{1,3} \wedge r_1 \wedge r_2 \wedge p_{2,1} \wedge \neg p_{2,2} \wedge \neg p_{2,3}) \vee$
   $\qquad\quad (\neg p_{1,1} \wedge p_{1,2} \wedge \neg p_{1,3} \wedge \neg r_1 \wedge r_2 \wedge p_{2,1} \wedge \neg p_{2,2} \wedge \neg p_{2,3}) \vee$
   $\qquad\quad (p_{1,1} \wedge \neg p_{1,2} \wedge \neg p_{1,3} \wedge r_1 \wedge \neg r_2 \wedge \neg p_{2,1} \wedge p_{2,2} \wedge \neg p_{2,3}).$

3. The second round of the **for** loop:

   $-\ \mathbf{SUC}(New) = (\neg p_{1,1} \wedge \neg p_{1,2} \wedge p_{1,3} \wedge \neg r_1 \wedge \neg r_2 \wedge p_{2,1} \wedge \neg p_{2,2} \wedge \neg p_{2,3}) \vee$
   $\qquad\qquad\qquad (\neg p_{1,1} \wedge p_{1,2} \wedge \neg p_{1,3} \wedge \neg r_1 \wedge \neg r_2 \wedge \neg p_{2,1} \wedge p_{2,2} \wedge \neg p_{2,3}) \vee$
   $\qquad\qquad\qquad (p_{1,1} \wedge \neg p_{1,2} \wedge \neg p_{1,3} \wedge \neg r_1 \wedge \neg r_2 \wedge \neg p_{2,1} \wedge \neg p_{2,2} \wedge p_{2,3}),$
   $-\ New = \mathbf{SUC}(New) - \mathbb{M} = \mathbf{SUC}(New),$
   $-\ \mathbb{M} = \mathbb{M} \vee New$
   $\qquad = (p_{1,1} \wedge \neg p_{1,2} \wedge \neg p_{1,3} \wedge r_1 \wedge r_2 \wedge p_{2,1} \wedge \neg p_{2,2} \wedge \neg p_{2,3}) \vee$
   $\qquad\quad (p_{1,1} \wedge \neg p_{1,2} \wedge \neg p_{1,3} \wedge r_1 \wedge \neg r_2 \wedge \neg p_{2,1} \wedge p_{2,2} \wedge \neg p_{2,3}) \vee$
   $\qquad\quad (p_{1,1} \wedge \neg p_{1,2} \wedge \neg p_{1,3} \wedge \neg r_1 \wedge \neg r_2 \wedge \neg p_{2,1} \wedge \neg p_{2,2} \wedge p_{2,3}) \vee$
   $\qquad\quad (\neg p_{1,1} \wedge p_{1,2} \wedge \neg p_{1,3} \wedge \neg r_1 \wedge r_2 \wedge p_{2,1} \wedge \neg p_{2,2} \wedge \neg p_{2,3}) \vee$
   $\qquad\quad (\neg p_{1,1} \wedge \neg p_{1,2} \wedge p_{1,3} \wedge \neg r_1 \wedge \neg r_2 \wedge p_{2,1} \wedge \neg p_{2,2} \wedge \neg p_{2,3}) \vee$
   $\qquad\quad (\neg p_{1,1} \wedge p_{1,2} \wedge \neg p_{1,3} \wedge \neg r_1 \wedge \neg r_2 \wedge \neg p_{2,1} \wedge p_{2,2} \wedge \neg p_{2,3}).$

4. The third round of the **for** loop:

   – **SUC**(New) $= p_{1,1} \wedge \neg p_{1,2} \wedge \neg p_{1,3} \wedge r_1 \wedge r_2 \wedge p_{2,1} \wedge \neg p_{2,2} \wedge \neg p_{2,3}$,
   – $New = $ **SUC**(New) $- \mathbb{M} = $ **false**,
   – $\mathbb{M} = \mathbb{M} \vee New = \mathbb{M}$.

5. The fourth round of the **for** loop is terminated due to $New = $ **false**.

5.3.1.4 Encoding All Edges of a Reachability Graph

We can also use an ROBDD to represent the directed edges of a reachability graph, without considering the transition names associated with these edges. Since such an edge is a pair of markings, we construct a pair of Boolean variables for each place such that half of these variables are used to represent the first marking and another half are used to represent the second marking.

Example 5.20 For the Petri net in Fig. 5.2, we construct 16 Boolean variables that are named as

$$r_1, \ r_1', \ r_2, \ r_2', \ p_{j,k} \ p_{j,k}', \quad \forall j \in \{1, 2\}, \ \forall k \in \{1, 2, 3\}.$$

These 8 directed edges as shown in Fig. 5.1a are represented by 8 Boolean functions as shown in Table 5.2. All such functions are combined with the disjunction operators and then we obtain the Boolean function corresponding to all edges. Here, we define the ordering of the 16 variables as follows:

$$p_{1,1} < p_{1,1}' < p_{1,2} < p_{1,2}' < p_{1,3} < p_{1,3}' < r_1 < r_1'$$

$$< r_2 < r_2' < p_{2,1} < p_{2,1}' < p_{2,2} < p_{2,2}' < p_{2,3} < p_{2,3}'$$

and then we obtain its ROBDD as shown in Fig. 5.10.

Since encoding edges doubles the amount of variables in comparison with encoding states and the amount of edges is usually much more than the amount of states, it is not a good way to encode both of them. If only states are encoded by an ROBDD in a model checking method, it will save much storage space. Fortunately, the structural characteristics of Petri nets can support this policy.

Table 5.2 Boolean Functions Representing the Edges of the Reachability Graph in Fig. 5.1a

Edges	Boolean functions															
	$P_{1,1}$	$P_{1,2}$	$P_{1,3}$	r_1	r_2	$P_{2,1}$	$P_{2,2}$	$P_{2,3}$	$P'_{1,1}$	$P'_{1,2}$	$P'_{1,3}$	r'_1	r'_2	$P'_{2,1}$	$P'_{2,2}$	$P'_{2,3}$
$s_0 \longrightarrow s_1$	1	0	0	1	1	1	0	0	0	1	0	0	1	1	0	0
$s_1 \longrightarrow s_2$	0	1	0	0	1	1	0	0	0	0	1	0	0	1	0	0
$s_2 \longrightarrow s_0$	0	0	1	0	0	1	0	0	1	0	0	1	1	1	0	0
$s_0 \longrightarrow s_3$	1	0	0	1	1	1	0	0	1	0	0	1	0	0	1	0
$s_3 \longrightarrow s_4$	1	0	0	1	0	0	1	0	1	0	0	0	0	0	0	1
$s_4 \longrightarrow s_0$	1	0	0	0	0	0	0	1	1	0	0	1	1	1	0	0
$s_1 \longrightarrow s_5$	0	1	0	0	1	1	0	0	0	1	0	0	0	0	1	0
$s_3 \longrightarrow s_5$	1	0	0	1	0	0	1	0	0	1	0	0	0	0	1	0

Notice: each function is the conjunction of the 16 variables or their negations: 1 corresponds to a variable itself and 0 corresponds to the negation of a variable

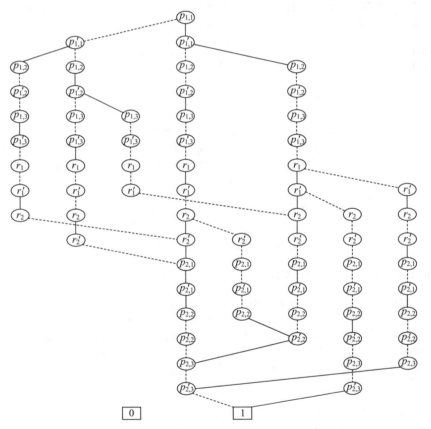

Fig. 5.10 The ROBDD representing the edges of the reachability graph in Fig. 5.1a. Here the edges connecting all non-terminal nodes with the terminal node v^0 are not drawn for simplicity

---

Algorithm 5.10: PRE$(\mathbb{M}, f)$: Generating the Predecessors of a Given Group of Markings.

Input: All reachable markings $\mathbb{M}$ and a group of reachable markings f of the Petri net
$\quad\quad\quad (P, T, F, M_0)$ encoded by an ROBDD;
Output: The predecessors of this group of reachable markings
1 $Pre \leftarrow$ **false**;
2 **for** *each* $t \in T$ **do**
3 $\quad$ $Enable \leftarrow \bigwedge_{p \in t^\bullet} p \wedge \bigwedge_{p \in {}^\bullet t \backslash t^\bullet} \neg p \wedge f$;
4 $\quad$ **if** $Enable \neq$ **false then**
5 $\quad\quad$ $\lfloor$ $Enable \leftarrow Enable[p \rightsquigarrow \neg p]_{p \in ({}^\bullet t \cup t^\bullet) \backslash ({}^\bullet t \cap t^\bullet)}$;
6 $\quad$ $\lfloor$ $Pre \leftarrow Pre \vee Enable$;
7 **return** $Pre \wedge \mathbb{M}$;

5.3.2   CTL Verification Based on All Reachable Markings Encoded by an ROBDD

5.3.2.1   Batch Generation of Predecessors

As shown in the algorithms of checking CTL formulae, one key factor is to compute the predecessors of a given group of markings. In the previous section, we have shown how to compute the successors of a given markings when they are represented by a Boolean function, which makes full use of the input and output of a transition.

Similarly, given a group of reachable markings of a Petri net, we can still use the structure of this Petri net to compute their predecessors. Computing their predecessors can be transferred into computing their successors in the inverse net. Every predecessor that we want is such a successor, but some successors computed by this way possibly are not predecessors that we want.

Example 5.21 For instance of the Petri net in Fig. 5.11a, it has two reachable markings: the initial one $\{p_2, p_3\}$ and the one $\{p_4\}$ reached by firing transition t_2 at the initial one. Obviously, the predecessor of the marking $\{p_4\}$ is exactly $\{p_2, p_3\}$. Its inverse net is shown in Fig. 5.11b. However, the marking $\{p_4\}$ of the inverse net has two successors $\{p_1, p_2\}$ and $\{p_2, p_3\}$, and only the latter is what we want.

How to remove these pseudo markings?

Fig. 5.11 An illustration of
Pseudo predecessors. (a) a
simple Petri net and (b) its
inverse net

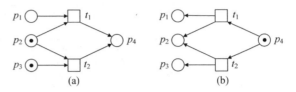

(a)   (b)

Obviously, if we have known all reachable markings, we can know that these
pseudo predecessors are not reachable in the original Petri net. Fortunately,
the operation of removing these pseudo ones can be easily implemented over
an ROBDD, i.e.,

$$Pre \wedge \mathbb{M}$$

where $\mathbb{M}$ is the ROBDD encoding all reachable markings of a Petri net and
Pre is all possible predecessors computed by the above idea for a given set of
markings. Algorithm 5.10 shows such a computing process. However, when
considering the enabling condition of a transition in the inverse net, we must
require that none of the output places of the transition has a token (unless a
place is both the input and the output of the transition), and otherwise such
a place has more than one token at the generated marking, which contradicts
the assumption that the Petri net we consider is safe. Therefore, as shown in
Line 3 of Algorithm 5.10, we have

$$Enable \leftarrow \left(\bigwedge_{p \in t^{\bullet}} p \right) \wedge \left(\bigwedge_{p \in {}^{\bullet}t \setminus t^{\bullet}} \neg p \right) \wedge f$$

that is different from the *Enable* in Algorithm 5.9.

Example 5.22 We set $\{p_1, p_2, p_3\}$ as the initial marking of the net in Fig. 5.11a.
Then it has two successors $\{p_3, p_4\}$ reached by firing t_1 and $\{p_1, p_4\}$ reached by
firing t_2. Obviously, this is a safe Petri net. Considering the marking $\{p_1, p_4\}$ and the
transition t_1 of the net in Fig. 5.11b, we obviously know that this marking is reached
by firing t_2 rather than t_1 in (a). If we only use the inputs of a transition as an enabling
condition, we have

$$p_4 \wedge (p_1 \wedge p_4) = p_1 \wedge p_4 \neq \textbf{false},$$

which means that using t_1 can compute a successor of $\{p_1, p_4\}$ but this is not what
we want. If we take the inputs and outputs of a transition as an enabling condition,
then we have

$$p_4 \wedge (\neg p_1 \wedge \neg p_2) \wedge (p_1 \wedge p_4) = \textbf{false},$$

which means that we cannot use t_1 to compute a successor of $\{p_1, p_4\}$ and this is
what we want.

Algorithm 5.11: SAT$(\mathbb{M}, \phi)$**: Computing the Satisfiable Marking Set for a Given CTL Formula through the ROBDD Encoding All Reachable Markings of a Given Petri Net.**

Input: The ROBDD $\mathbb{M}$ encoding all reachable markings of a given Petri net and a CTL formula ϕ;

Output: The satisfiable set **SAT**$(\mathbb{M}, \phi)$;

1 **if** $\phi = $ **true then**
2   **return** $\mathbb{M}$;

3 **if** $\phi = p$ **then**
4   **return** $p \wedge \mathbb{M}$;

5 **if** $\phi = \neg\phi_1$ **then**
6   **return** $\mathbb{M} - $ **SAT**$(\mathbb{M}, \phi_1)$;

7 **if** $\phi = \phi_1 \wedge \phi_2$ **then**
8   **return SAT**$(\mathbb{M}, \phi_1) \wedge $ **SAT**$(\mathbb{M}, \phi_2)$;

9 **if** $\phi = $ **EX** ϕ_1 **then**
10   **return PRE**$(\mathbb{M}, $ **SAT**$(\mathbb{M}, \phi_1))$;

11 **if** $\phi = $ **EG** ϕ_1 **then**
12    $X \leftarrow $ **SAT**$(\mathbb{M}, \phi_1)$;
13    $Z \leftarrow X \wedge (\mathbb{M} - ((\bigvee_{t \in T} \bigwedge_{p \in {}^\bullet t} p) \wedge \mathbb{M}))$;
14    $Y \leftarrow (\textbf{PRE}(\mathbb{M}, X) \wedge X) \vee Z$;
15   **while** $X \neq Y$ **do**
16      $X \leftarrow Y$;
17      $Y \leftarrow (\textbf{PRE}(\mathbb{M}, X) \wedge X) \vee Z$;
18   **return** Y;

19 **if** $\phi = $ **E**$(\phi_1$ **U** $\phi_2)$ **then**
20    $X \leftarrow $ **SAT**$(\mathbb{M}, \phi_1)$;
21    $Y \leftarrow $ **SAT**$(\mathbb{M}, \phi_2)$;
22   **while** $X \neq $ **false do**
23      $Z \leftarrow X \wedge $ **PRE**$(\mathbb{M}, Y)$;
24      $X \leftarrow X - Z$;
25      $Y \leftarrow Y \vee Z$;
26   **return** Y;

5.3.2.2 Verification Algorithm

Based on **PRE** as well as the algorithms of checking CTL formulae presented in the previous section, we easily write Algorithm 5.11 to compute the satisfiable marking set of a CTL formula through the ROBDD representing all reachable markings. Here, $\mathbb{M}$ denotes the ROBDD encoding all reachable markings of a given Petri net.

Before introducing the computation of the satisfiable set of the formula $\phi = $ **EG** ϕ_1, we first introduce the computation of **deadlock** in a given Petri net, because these deadlocks are used in the computation of the satisfiable marking set of a formula of **EG**.

We know that the Boolean expression

$$\left(\bigwedge_{p \in {}^\bullet t} p \right) \wedge \mathbb{M}$$

represents all reachable markings enabling the transition t.

As a result, the Boolean expression

$$\left(\bigvee_{t \in T} \bigwedge_{p \in {}^\bullet t} p \right) \wedge \mathbb{M}$$

represents all reachable markings except deadlocks.

Therefore, all reachable deadlocks can be represented by the following expression:

$$\mathbb{M} - \left(\left(\bigvee_{t \in T} \bigwedge_{p \in {}^\bullet t} p \right) \wedge \mathbb{M} \right).$$

Lines 11–18 in Algorithm 5.11 is to compute the satisfiable set of $\phi = \mathbf{EG}\,\phi_1$. It first computes the satisfiable set of the formula ϕ_1 (Line 12) and then computes those deadlocks satisfying ϕ_1 (Line 13). Next, these deadlocks are considered every time (Lines 14 and 17), which means that each of them can be thought of as a predecessor of itself. The **while** loop is iteratively to compute the predecessors of those markings satisfying ϕ_1 and requires that these predecessors also satisfy ϕ_1, until the computed predecessors are not changed any more. The computations of the satisfiable sets of **EX** and **EU** are easily understood according to Algorithms 5.2 and 5.4 and thus omitted here.

Example 5.23 We check the formula $\mathbf{EG}(p_{1,2} \vee p_{2,1})$ in the Petri net in Fig. 5.2 whose reachability graph is shown in Fig. 5.1a. First, we obtain

$-\ X = \mathbf{SAT}(p_{1,2} \vee p_{2,1})$
$$= (p_{1,1} \wedge \neg p_{1,2} \wedge \neg p_{1,3} \wedge r_1 \wedge r_2 \wedge p_{2,1} \wedge \neg p_{2,2} \wedge \neg p_{2,3}) \vee$$
$$(\neg p_{1,1} \wedge p_{1,2} \wedge \neg p_{1,3} \wedge \neg r_1 \wedge r_2 \wedge p_{2,1} \wedge \neg p_{2,2} \wedge \neg p_{2,3}) \vee$$
$$(\neg p_{1,1} \wedge \neg p_{1,2} \wedge p_{1,3} \wedge \neg r_1 \wedge \neg r_2 \wedge p_{2,1} \wedge \neg p_{2,2} \wedge \neg p_{2,3}) \vee$$
$$(\neg p_{1,1} \wedge p_{1,2} \wedge \neg p_{1,3} \wedge \neg r_1 \wedge \neg r_2 \wedge \neg p_{2,1} \wedge p_{2,2} \wedge \neg p_{2,3})$$

in which the deadlock is

$-\ Z = \neg p_{1,1} \wedge p_{1,2} \wedge \neg p_{1,3} \wedge \neg r_1 \wedge \neg r_2 \wedge \neg p_{2,1} \wedge p_{2,2} \wedge \neg p_{2,3}.$

According to the algorithm of computing predecessors, we have

$$
\begin{aligned}
- \ \mathbf{PRE}(\mathbb{M}, X) = \ & (p_{1,1} \wedge \neg p_{1,2} \wedge \neg p_{1,3} \wedge r_1 \wedge r_2 \wedge p_{2,1} \wedge \neg p_{2,2} \wedge \neg p_{2,3}) \vee \\
& (p_{1,1} \wedge \neg p_{1,2} \wedge \neg p_{1,3} \wedge r_1 \wedge \neg r_2 \wedge \neg p_{2,1} \wedge p_{2,2} \wedge \neg p_{2,3}) \vee \\
& (p_{1,1} \wedge \neg p_{1,2} \wedge \neg p_{1,3} \wedge \neg r_1 \wedge \neg r_2 \wedge \neg p_{2,1} \wedge \neg p_{2,2} \wedge p_{2,3}) \vee \\
& (\neg p_{1,1} \wedge p_{1,2} \wedge \neg p_{1,3} \wedge \neg r_1 \wedge r_2 \wedge p_{2,1} \wedge \neg p_{2,2} \wedge \neg p_{2,3}) \vee \\
& (\neg p_{1,1} \wedge \neg p_{1,2} \wedge p_{1,3} \wedge \neg r_1 \wedge \neg r_2 \wedge p_{2,1} \wedge \neg p_{2,2} \wedge \neg p_{2,3}).
\end{aligned}
$$

Then, we have

$$
\begin{aligned}
- \ Y = \ & (\mathbf{PRE}(\mathbb{M}, X) \wedge X) \vee Z \\
= \ & (p_{1,1} \wedge \neg p_{1,2} \wedge \neg p_{1,3} \wedge r_1 \wedge r_2 \wedge p_{2,1} \wedge \neg p_{2,2} \wedge \neg p_{2,3}) \vee \\
& (\neg p_{1,1} \wedge p_{1,2} \wedge \neg p_{1,3} \wedge \neg r_1 \wedge r_2 \wedge p_{2,1} \wedge \neg p_{2,2} \wedge \neg p_{2,3}) \vee \\
& (\neg p_{1,1} \wedge \neg p_{1,2} \wedge p_{1,3} \wedge \neg r_1 \wedge \neg r_2 \wedge p_{2,1} \wedge \neg p_{2,2} \wedge \neg p_{2,3}) \vee \\
& (\neg p_{1,1} \wedge p_{1,2} \wedge \neg p_{1,3} \wedge \neg r_1 \wedge \neg r_2 \wedge \neg p_{2,1} \wedge p_{2,2} \wedge \neg p_{2,3}).
\end{aligned}
$$

Since X is equal to Y, the loop body of **while** is not executed. Hence, we have

$$
\begin{aligned}
- \ \mathbf{EG}(p_{1,2} \vee p_{2,1}) = \ & (p_{1,1} \wedge \neg p_{1,2} \wedge \neg p_{1,3} \wedge r_1 \wedge r_2 \wedge p_{2,1} \wedge \neg p_{2,2} \wedge \neg p_{2,3}) \vee \\
& (\neg p_{1,1} \wedge p_{1,2} \wedge \neg p_{1,3} \wedge \neg r_1 \wedge r_2 \wedge p_{2,1} \wedge \neg p_{2,2} \wedge \neg p_{2,3}) \vee \\
& (\neg p_{1,1} \wedge \neg p_{1,2} \wedge p_{1,3} \wedge \neg r_1 \wedge \neg r_2 \wedge p_{2,1} \wedge \neg p_{2,2} \wedge \neg p_{2,3}) \vee \\
& (\neg p_{1,1} \wedge p_{1,2} \wedge \neg p_{1,3} \wedge \neg r_1 \wedge \neg r_2 \wedge \neg p_{2,1} \wedge p_{2,2} \wedge \neg p_{2,3})
\end{aligned}
$$

which is coincided with the fact.

5.4 Application

Here we show how to use CTL formulae to describe some important properties of safe Petri nets. Deadlock and fairness have been described above.

Reachability

Given a marking M, deciding if it is reachable can be transferred into deciding if the following formula is satisfiable:

$$
\mathbf{EF}\left(\bigwedge_{\substack{p \in P \\ M(p)=1}} p \wedge \bigwedge_{\substack{p \in P \\ M(p)=0}} \neg p \right).
$$

Livelock

Livelock's definition is related to a final marking. Hence, given a final marking M_d, deciding if there is a livelock can be transferred into deciding if the following formula is satisfiable:

$$
\mathbf{EG}\left(\neg \left(\bigwedge_{\substack{p \in P \\ M_d(p)=1}} p \wedge \bigwedge_{\substack{p \in P \\ M_d(p)=0}} \neg p \right) \wedge \bigvee_{t \in T} \bigwedge_{p \in {}^{\bullet}t} p \right).
$$

In the sub-formula of **EG**, the first half represents the final marking is not reached, and the last half represents there exists at least one enabled transition.

Liveness

The liveness of a transition t can be represented by the following formula:

$$\mathbf{AG}\left(\mathbf{EF}\bigwedge_{p\in\bullet t}p\right).$$

Hence, liveness can be represented by the following formula:

$$\bigwedge_{t\in T}\left(\mathbf{AG}\left(\mathbf{EF}\bigwedge_{p\in\bullet t}p\right)\right).$$

Soundness

Weak soundness is only related to the final marking and can be represented by the following formula:

$$\mathbf{AF}\left(\bigwedge_{\substack{p\in P\\M_d(p)=1}}p\wedge\bigwedge_{\substack{p\in P\\M_d(p)=0}}\neg p\right).$$

Soundness also requires that each transition has a potential chance to be fired, and thus it can be represented by the following formula:

$$\left(\mathbf{AF}\left(\bigwedge_{\substack{p\in P\\M_d(p)=1}}p\wedge\bigwedge_{\substack{p\in P\\M_d(p)=0}}\neg p\right)\right)\wedge\left(\bigwedge_{t\in T}\mathbf{EF}\left(\bigwedge_{p\in\bullet t}p\right)\right).$$

Collaborative-ness

We denote $P_I=\{i_1,\ldots,i_m\}$ and $P_O=\{o_1,\ldots,o_m\}$. The weak collaborative-ness can be represented by the following formula:

$$\bigwedge_{i_k\in P_I}\mathbf{AG}\left(\neg i_k\Rightarrow\mathbf{AF}\left(o_k\wedge\bigwedge_{p\in P\backslash(P_I\cup P_O)}\neg p\right)\right).$$

Similar to the formula representing soundness, we additionally take into account transitions on the basis of the above formula and thus obtain the formula of collaborative-ness.

5.5 Summery and Further Reading

This chapter introduces the basic knowledge of CTL, ROBDD as well as the CTL verifications based on Petri nets and ROBDDs. Some related algorithms are presented that can reflect those basic principles but not necessarily very efficient (e.g. Algorithms 5.5 and 5.7). We have pointed out some related references in which more details can be found by readers. The following problems are worthy being explored further.

Since a variable ordering determines the scale of an ROBDD, how to obtain a good variable ordering is an important issue [16–19], including how to make full use of the structure of a Petri net to obtain an ordering [11, 20, 21] and how to dynamically change an ordering in the procedure of generating an ROBDD [10, 22, 23]. In [11], we propose a method to generate a variable ordering on the basis of both a Petri net structure and a part of its reachable markings. This method is very suitable for modular loosely-coupled Petri nets, and experiments on the dining philosophers problem illustrate that it can check the deadlock of the case with 5000 philosophers in 12 h on a PC. The compressed markings are about 10^{3300}, which is amazing!

As stated in Chap. 1, a finite complete prefix of the unfolding of a bounded Petri net can also represent all reachable states, and its scale is very small when the Petri net has a number of concurrent transitions. It has been used as a technique of compressing states of Petri nets as well as checking some temporal logics [24–32], but more efficient algorithms may be explored further.

When a system's states are so many that all of them cannot be generated or encoded within acceptable time, a so-called *bounded model checking* can be utilised. Readers can study the related knowledge in [4, 33–35].

References

1. Clarke, E.M., Emerson, E.A.: Design and synthesis of synchronization skeletons using branching time temporal logic. In: *Proceedings of the IBM Workshop on Logics of Programs*, pp. 52–7I (1981)
2. Queille, J.P., Sifakis, J.: Specification and verification of concurrent systems in CESAR. In: *the 5th International Symposium on Programming*, pp. 337–351 (1982)
3. Clarke, E.M., Oma Grumberg, Jr., Peled, D.A.: *Model Checking*. The MIT Press, London (1999)
4. Baier, C., Katoen, J.P.: *Principles of Model Checking*. The MIT Press, London (2008)
5. Bønneland, F., Dyhr, J., Jensen, P.G., Johannsen, M., Srba, J.: Simplification of CTL formulae for efficient model checking of Petri nets. In: *the 39th International Conference on Application and Theory of Petri Nets and Concurrency*. pp. 143–163 (2018)
6. Bryant, R.E.: Graph-based algorithms for Boolean function manipulation. *IEEE Transactions on Computers* 8: 677–691 (1986)
7. Bryant, R.E.: Symbolic Boolean manipulation with ordered binary-decision diagrams. *ACM Computing Surveys* 24: 293–318 (1992)
8. Andersen, H.R.: An introduction to binary decision diagrams. *Technical Report*, Technical University of Denmark (1998)

 9. Somenzi, F.: CUDD: CU decision diagram package-release 2.5.1. http://vlsi.colorado.edu/
    fabio/CUDD (2021)
10. Rudell, R.: Dynamic variable ordering for ordered binary decision diagrams. In: *the IEEE/ACM
    International Conference on Computer-aided Design*, pp. 51–63 (1993)
11. He, L.F., Liu, G.J.: Petri net based CTL model checking: using a new method to construct
    OBDD variable order. In: *the 15th International Symposium on Theoretical Aspects of Software
    Engineering*, pp. 159–166 (2021)
12. Heiner, M., Rohr, C., Schwarick, M.: MARCIE: model checking and reachability analysis
    done efficiently. In: *the 24th International Conference on Applications and Theory of Petri
    Nets and Concurrency*, pp. 389–399 (2013)
13. Amparore, E.G., Beccuti, M., Donatelli, S.: (Stochastic) model checking in GreatSPN. In: *the
    35th International Conference on Applications and Theory of Petri Nets and Concurrency*, pp.
    354–363 (2014)
14. Kant, G., Laarman, A., Meijer, J., Pol, J.V.D., Blom, S., Dijk, T.V.: LTSmin: high-performance
    language-independent model checking. In: *International Conference on Tools and Algorithms
    for the Construction and Analysis of Systems*, pp. 692–707 (2015)
15. Thierry-Mieg, Y.: Symbolic model-checking using ITS-tools. In: *International Conference on
    Tools and Algorithms for the Construction and Analysis of Systems*, pp. 231–237 (2015)
16. Chung, P.Y., Hajj, I.M., Patel, J.H.: Efficient variable ordering heuristics for shared ROBDD.
    In: *1993 IEEE International Symposium on Circuits and Systems*, pp. 1690-1693 (1993)
17. Pastor, E., Roig, O., Cortadella, J., Badia, R.M.: Petri net analysis using boolean manipulation.
    In: *the 15th International Conference on Application and Theory of Petri Nets*, pp. 416–435
    (1994)
18. Varma, C.: An enhanced algorithm for variable reordering in binary decision diagrams. In: *the
    9th International Conference on Computing, Communication and Networking Technologies*,
    pp. 1–4 (2018)
19. Newton, J., Verna, D.: A theoretical and numerical analysis of the worst-case size of reduced
    ordered binary decision diagrams. *ACM Transactions on Computational Logic* 20: 1–36 (2019)
20. Noack, A.: A ZBDD package for efficient model checking of Petri nets. Technical Report,
    Department of Computer Science, Brandenburg Technology University Cottbus, Germany
    (1999) (In German)
21. Tovchigrechko, A.: Model checking using interval decision diagrams. Ph.D. Thesis, Depart-
    ment of Computer Science, Brandenburg Technology University Cottbus, Germany (2008)
22. Lomuscio, A., Qu, H., Raimondi, F.: MCMAS: An open-source model checker for the verifica-
    tion of multi-agent systems. *International Journal on Software Tools for Technology Transfer*
    19: 9–30 (2017)
23. Li, J., Yang, Y., Huo, G., Huang, G., Jin, Y.: New bidirectional fast BDD dynamic reordering
    algorithm. In: *the 15th IEEE International Conference on Solid-State and Integrated Circuit
    Technology*, pp. 1–3 (2020)
24. McMillan, K.L.: *Symbolic Model Checking*. Kluwer Academic Publishers, Dordrecht, The
    Netherlands (1993)
25. Khomenko, V.: Model checking based on prefixes of Petri net unfoldings. Ph.D. Dissertation,
    School of Computer Science, Newcastle University (2003)
26. Esparza, J., Heljanko, K.: *Unfoldings: A Partial-Order Approach to Model Checking*. Springer-
    Verlag, Berlin Heidelberg (2008)
27. Liu, G.J., Zhang, K., Jiang, C.J.: Deciding the deadlock and livelock in a petri net with a target
    marking based on its basic unfolding. In: *the 16th International Conference on Algorithms
    and Architectures for Parallel Processing*, pp. 98–105 (2016)
28. Zhang, K., Liu, G.J., Xiang, D.M.: BUCKER: A basic unfolding based checker for soundness
    of workflow systems. In: *the 14th IEEE International Conference on Networking, Sensing and
    Control*, pp. 611–616 (2017)
29. Dong, L.L., Liu, G.J., Xiang, D.M.: Verifying CTL with unfoldings of Petri nets. In: *the 18th
    International Conference on Algorithms and Architectures for Parallel Processing*, pp. 47–61
    (2018)

30. Dong, L.L., Liu, G.J., Xiang, D.M.: BUCKER 2.0: An Unfolding Based Checker for CTL. In: *the 16th IEEE International Conference on Networking, Sensing and Control*, pp. 144–149 (2019)
31. Liu, G.J.: *The Primary Unfoldings of Petri Nets: A Model Checking Method for Concurrent Systems*. China Science Press, Beijing, China (2020) (in Chinese)
32. Finkbeiner, B., Gieseking, M., Hecking-Harbusch, J., Olderog, E.R.: Model checking branching properties on Petri nets with transits. In: *International Symposium on Automated Technology for Verification and Analysis*, pp. 394–410 (2020)
33. Clarke, E., Biere, A., Raimi, R., Zhu, Y.: Bounded model checking using satisfiability solving. *Formal methods in system design* 19: 7–34 (2001)
34. Finkbeiner, B., Gieseking, M., Hecking-Harbusch, J., Olderog, E.R.: AdamMC: A model checker for Petri nets with transits against Flow-LTL. In: *International Conference on Computer Aided Verification*, pp. 64–76 (2020)
35. Amat, N., Berthomieu, B., Dal Zilio, S.: On the combination of polyhedral abstraction and SMT-based model checking for Petri nets. In: *the 42nd International Conference on Applications and Theory of Petri Nets and Concurrency*, pp. 164–185 (2021)

Chapter 6
Knowledge-Oriented Petri Nets and Computation Tree Logic of Knowledge

6.1 Knowledge-Oriented Petri Nets Modelling Privacy-Critical Multi-agent Systems

This section first introduces the definition of knowledge-oriented Petri nets. And then, equivalence relations w.r.t. the epistemic evolution of each agent are described. Readers can find more details in [1–6].

6.1.1 Knowledge-Oriented Petri Nets

Definition 6.1 (*Knowledge-oriented Petri Net*) A *knowledge-oriented net* is a 6-tuple $N^{\mathfrak{k}} \triangleq (P_U, P_K, T, F, \mathcal{A}, L)$ where

1. $(P_U \cup P_K, T, F)$ is a net where

   a. P_U is a finite set of *non-knowledge places*,
   b. P_K is a finite set of *knowledge places*, and
   c. $P_U \cap P_K = \emptyset$,

2. $\mathcal{A}$ is a finite set of *agent names*, and
3. $L: P_K \rightarrow 2^{\mathcal{A}}$ is a *labeling function*.

A *knowledge-oriented Petri net* is a knowledge-oriented net $N^{\mathfrak{k}}$ with an initial marking M_0 and denoted as $(N^{\mathfrak{k}}, M_0)$.

A knowledge-oriented Petri net is actually a Petri net in which some places (defined by P_K) are used to represent different knowledge and the labelling function L indicates which agents *know* which pieces of knowledge. We use a key transport protocol with public-key encryption [7] to illustrate this definition.

Example 6.1 The knowledge-oriented Petri net in Fig. 6.1 models this protocol where hollow circles stand for non-knowledge places and solid ones stand for

© The Author(s), under exclusive license to Springer Nature Singapore Pte Ltd. 2022
G. Liu, *Petri Nets*,
https://doi.org/10.1007/978-981-19-6309-4_6

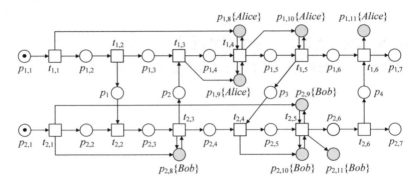

Fig. 6.1 The knowledge-oriented Petri net modelling the key transport protocol with public-key encryption

knowledge places. In this example, there are two agents Alice and Bob (i.e., $\mathcal{A} = \{Alice, Bob\}$). First, Alice chooses her private password ($t_{1,1}$). Bob chooses his public key and private key, respectively ($t_{2,1}$). Then Alice sends Bob a request asking for his public key ($t_{1,2}$). Once receiving the request ($t_{2,2}$), Bob delivers his public key to her ($t_{2,3}$). After Alice receives the public key ($t_{1,3}$), she uses it to encrypt her password ($t_{1,4}$) and sends the encrypted password to Bob ($t_{1,5}$). After receiving the encrypted password ($t_{2,4}$), Bob uses his private key to decrypt it ($t_{2,5}$). Finally, Bob sends an acknowledgement to Alice ($t_{2,6}$).

A token in the knowledge place $p_{1,8}$ means a piece of knowledge that Alice has set up a password. The password will be encrypted later, which is represented by a self-loop between the place and the transition $t_{1,4}$, whereas this self-loop means that this password is encrypted but does not result in the loss of this piece of knowledge. A token in the knowledge place $p_{2,8}$ (respectively, $p_{2,9}$) means a piece of knowledge that Bob has set up a public key (respectively, a private key). A token in $p_{1,9}$ means a piece of knowledge that Alice has received the public key. A token in $p_{1,10}$ (respectively, $p_{2,10}$) means a piece of knowledge that Alice (respectively, Bob) has got the encrypted password. A token in $p_{2,11}$ means a piece of knowledge that Bob has got the password. A token in $p_{1,11}$ means a piece of knowledge that Alice has got an acknowledgement.

The places p_1, p_2, p_3, and p_4 stand for the transmission of these messages through a channel. And later we will show that if the channel is not secure, then the password can be obtained by an attacker but Alice does not know this.

The underlying Petri net w.r.t. P_U models the interactive process of these agents. We use $p \in P_K$ to represent a piece of knowledge obtained by a set of agents defined by $L(p)$. In the above example, there is no common knowledge, i.e., every piece of knowledge is known by exactly one agent. Later, we will show another example in which a piece of knowledge can be known by multiple agents.

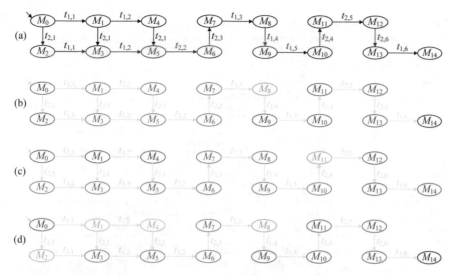

Fig. 6.2 a The reachability graph of the knowledge-oriented Petri net in Fig. 6.1; **b** the equivalence relation w.r.t. the agent *Alice*, i.e., $\sim_{Alice}$, including 5 equivalent classes; **c** the equivalence relation w.r.t. the agent *Bob*, i.e., $\sim_{Bob}$, including 4 equivalent classes; and **d** the equivalence relation ($\sim_{Alice}$ $\cap \sim_{Bob}$), including 9 equivalent classes. The markings in the same colour form an equivalent class and the information of each marking is listed in Table 6.1

> Now that a knowledge-oriented Petri net is a Petri net, the enabling and firing rules of its transitions are unchanged. The difference between a knowledge-oriented Petri net and a Petri net lies in some places that are viewed as different knowledge in the former, but it does not change the enabling and firing rules of transitions.

Therefore, given a knowledge-oriented Petri net, we can construct its reachability graph by which the interactive behaviours of agents can be represented, e.g. Fig. 6.2a shows the reachability graph of the knowledge-oriented Petri net in Fig. 6.1 where the distributions of tokens of all markings are listed in Table 6.1.

Similar to a Petri net, we use $R(N^{\varepsilon}, M_0)$ to represent the set of all reachable markings of (N^{ε}, M_0). In this chapter, we assume that all knowledge-oriented Petri nets discussed are safe.

However, how to describe the epistemic evolution of each agent? A method will be introduced in what follows.

Table 6.1 The token distribution of each marking in Fig. 6.2

Markings' names	The distributions of tokens and knowledge owned by agents
M_0	$\{p_{1,1}, p_{2,1}\}$
M_1	$\{p_{1,2}, \boldsymbol{p_{1,8}}, p_{2,1}\}$
M_2	$\{p_{1,1}, p_{2,2}, \boldsymbol{p_{2,8}}, \boldsymbol{p_{2,9}}\}$
M_3	$\{p_{1,2}, \boldsymbol{p_{1,8}}, p_{2,2}, \boldsymbol{p_{2,8}}, \boldsymbol{p_{2,9}}\}$
M_4	$\{p_{1,3}, \boldsymbol{p_{1,8}}, p_{2,1}, p_1\}$
M_5	$\{p_{1,3}, \boldsymbol{p_{1,8}}, p_{2,2}, \boldsymbol{p_{2,8}}, \boldsymbol{p_{2,9}}, p_1\}$
M_6	$\{p_{1,3}, \boldsymbol{p_{1,8}}, p_{2,3}, \boldsymbol{p_{2,8}}, \boldsymbol{p_{2,9}}\}$
M_7	$\{p_{1,3}, \boldsymbol{p_{1,8}}, p_{2,4}, \boldsymbol{p_{2,8}}, \boldsymbol{p_{2,9}}, p_2\}$
M_8	$\{p_{1,4}, \boldsymbol{p_{1,8}}, \boldsymbol{p_{1,9}}, p_{2,4}, \boldsymbol{p_{2,8}}, \boldsymbol{p_{2,9}}\}$
M_9	$\{p_{1,5}, \boldsymbol{p_{1,8}}, \boldsymbol{p_{1,9}}, \boldsymbol{p_{1,10}}, p_{2,4}, \boldsymbol{p_{2,8}}, \boldsymbol{p_{2,9}}\}$
M_{10}	$\{p_{1,6}, \boldsymbol{p_{1,8}}, \boldsymbol{p_{1,9}}, \boldsymbol{p_{1,10}}, p_{2,4}, \boldsymbol{p_{2,8}}, \boldsymbol{p_{2,9}}, p_3\}$
M_{11}	$\{p_{1,6}, \boldsymbol{p_{1,8}}, \boldsymbol{p_{1,9}}, \boldsymbol{p_{1,10}}, p_{2,5}, \boldsymbol{p_{2,8}}, \boldsymbol{p_{2,9}}, \boldsymbol{p_{2,10}}\}$
M_{12}	$\{p_{1,6}, \boldsymbol{p_{1,8}}, \boldsymbol{p_{1,9}}, \boldsymbol{p_{1,10}}, p_{2,6}, \boldsymbol{p_{2,8}}, \boldsymbol{p_{2,9}}, \boldsymbol{p_{2,10}}, \boldsymbol{p_{2,11}}\}$
M_{13}	$\{p_{1,6}, \boldsymbol{p_{1,8}}, \boldsymbol{p_{1,9}}, \boldsymbol{p_{1,10}}, p_{2,7}, \boldsymbol{p_{2,8}}, \boldsymbol{p_{2,9}}, \boldsymbol{p_{2,10}}, \boldsymbol{p_{2,11}}, p_4\}$
M_{14}	$\{p_{1,7}, \boldsymbol{p_{1,8}}, \boldsymbol{p_{1,9}}, \boldsymbol{p_{1,10}}, \boldsymbol{p_{1,11}} p_{2,7}, \boldsymbol{p_{2,8}}, \boldsymbol{p_{2,9}}, \boldsymbol{p_{2,10}}, \boldsymbol{p_{2,11}}\}$

6.1.2   Reachability Graph with Equivalence Relations

6.1.2.1   Equivalence Relations Based on Identical Knowledge

If an agent owns the same knowledge at two global states of a knowledge-oriented Petri net, the two states should be viewed as identity from the aspect of this agent's knowledge. In consequence, we can define an equivalence relation on the reachable marking set for every agent. For convenience, we first present some notations.

Given a knowledge-oriented Petri net $(N^{\mathfrak{k}}, M_0) = (P_U, P_K, T, F, \mathcal{A}, L, M_0)$ and an agent $\mathbf{a} \in \mathcal{A}$, we use $P_{\mathbf{a}}$ to represent those knowledge places w.r.t. $\mathbf{a}$, i.e.,

$$P_{\mathbf{a}} \triangleq \{p \in P_K \mid \mathbf{a} \in L(p)\}.$$

Similarly, given a group of agents $\Gamma \subseteq \mathcal{A}$, we use P_{Γ} to represent those knowledge places w.r.t. agents in Γ, i.e.,

$$P_{\Gamma} \triangleq \{p \in P_K \mid \exists \mathbf{a} \in \Gamma : \mathbf{a} \in L(p)\} = \bigcup_{\mathbf{a} \in \Gamma} P_{\mathbf{a}}.$$

Given a marking $M \in R(N^{\mathfrak{k}}, M_0)$, we use $||M||$ to represent the set of places marked at M,[1] i.e.,

---

[1] Since we assume that all discussed KPNs are safe, we have $||M|| = M$ when we use a (multi-)set to represent a marking. But for readability, we use $||M||$ to represent the support of M.

$$||M|| \triangleq \{p \in P \mid M(p) > 0\}.$$

Given a marking $M \in R(N^{\mathfrak{k}}, M_0)$ and an agent $\mathbf{a} \in \mathcal{A}$, we use $||M||_\mathbf{a}$ to represent the knowledge owned by $\mathbf{a}$ at M, i.e.,

$$||M||_\mathbf{a} \triangleq ||M|| \cap P_\mathbf{a} = \{p \in P_\mathbf{a} \mid M(p) > 0\}.$$

Similarly, given a group of agents $\Gamma \subseteq \mathcal{A}$, we use $||M||_\Gamma$ to represent the knowledge owned by the agents in Γ at M, i.e.,

$$||M||_\Gamma \triangleq ||M|| \cap P_\Gamma = \{p \in P_\Gamma \mid M(p) > 0\}.$$

Example 6.2 For the reachable marking M_9 in Table 6.1, we have

- $||M_9||_{Alice} = \{p_{1,8}, \ p_{1,9}, \ p_{1,10}\}$,
- $||M_9||_{\{Alice,Bob\}} = \{p_{1,8}, \ p_{1,9}, \ p_{1,10}, \ p_{2,8}, \ p_{2,9}\}$.

Now, for each agent $\mathbf{a} \in \mathcal{A}$ of $(N^{\mathfrak{k}}, M_0)$, we define an equivalence relation $\sim_\mathbf{a} \subseteq (R(N^{\mathfrak{k}}, M_0) \times R(N^{\mathfrak{k}}, M_0))$ w.r.t. the knowledge of $\mathbf{a}$ as follows:

$$\sim_\mathbf{a} \triangleq \{(M, M') \mid M \in R(N^{\mathfrak{k}}, M_0), M' \in R(N^{\mathfrak{k}}, M_0), ||M||_\mathbf{a} = ||M'||_\mathbf{a}\}.$$

Usually, $(M, M') \in \sim_\mathbf{a}$ is written as $M \sim_\mathbf{a} M'$.

Obviously, this binary relation is reflexive, symmetric, and transitive, i.e., it is an *equivalence relation*. $M \sim_\mathbf{a} M'$ means that the agent $\mathbf{a}$ owns the identical knowledge at M and M'. Consequently, the equivalence relation $\sim_\mathbf{a}$ divides the reachable set into a group of equivalent classes.

Example 6.3 The equivalence relation $\sim_{Alice}$ divides the reachable set listed in Table 6.1 into 5 equivalent classes:

$$\{M_0, M_2\}, \{M_1, M_3, M_4, M_5, M_6, M_7\}, \{M_8\}, \{M_9, M_{10}, M_{11}, M_{12}, M_{13}\}, \text{ and } \{M_{14}\},$$

as shown in Fig. 6.2b. Figure 6.2c illustrates the equivalence relation $\sim_{Bob}$, including 4 equivalent classes. In these figures, those markings in the same colour form an equivalent class.

6.1.2.2   Reachability Graph with Equivalence Relations and Operations of Equivalence Relations

Given an equivalence relation $\sim_a$ on the reachable set and a reachable marking M, we use the following notation to represent the equivalent class including M:

$$\widehat{M_{\sim_a}} \triangleq \{M' \mid M' \sim_a M\}.$$

Based on these equivalence relations, we define the *reachability graph with equivalence relations* of a knowledge-oriented Petri net as follows:

$$RGER(N^\natural, M_0, \sim) \triangleq (R(N^\natural, M_0), E, \sim_{a_1}, \ldots, \sim_{a_k}),$$

where $\mathcal{A} = \{a_1, \ldots, a_k\}$ and $(R(N^\natural, M_0), E)$ is its reachability graph.

Given two equivalence relations $\sim_a$ and $\sim_b$ in $RGER(N^\natural, M_0, \sim)$, three new binary relations $(\sim_a \cup \sim_b)$, $(\sim_a \cap \sim_b)$, and $(\sim_a \cup \sim_b)^+$, are defined, respectively, as follows:

- $M(\sim_a \cap \sim_b)M'$ if and only if $M \sim_a M'$ and $M \sim_b M'$,
- $M(\sim_a \cup \sim_b)M'$ if and only if $M \sim_a M'$ or $M \sim_b M'$, and
- $M(\sim_a \cup \sim_b)^+M'$ if and only if $M = M'$ or there exists $M_1, M_2, \ldots, M_n \in R(N^\natural, M_0)$ such that

$$M(\sim_a \cup \sim_b)M_1, M_1(\sim_a \cup \sim_b)M_2, \ldots, \text{ and } M_n(\sim_a \cup \sim_b)M'.$$

Based on the above operations, we can expand them to a group of equivalence relations, i.e.,

$$\left(\bigcap_{a \in \Gamma} \sim_a\right), \quad \left(\bigcup_{a \in \Gamma} \sim_a\right), \quad \text{and} \quad \left(\bigcup_{a \in \Gamma} \sim_a\right)^+$$

where $\Gamma \subseteq \mathcal{A}$ and $|\Gamma| \geq 1$. Notice, these three relations are all $\sim_a$ when $|\Gamma| = 1$. For simplicity, the above three notations are abbreviated to $(\bigcap_\Gamma \sim)$, $(\bigcup_\Gamma \sim)$, and $(\bigcup_\Gamma \sim)^+$, respectively.

According to the basic principles of equivalence relation [8], we have the following conclusions:

Property 6.1 $(\bigcap_\Gamma \sim)$ *and* $(\bigcup_\Gamma \sim)^+$ *are still equivalence relations when* $|\Gamma| \geq 1$, *but* $(\bigcup_\Gamma \sim)$ *is not necessarily so when* $|\Gamma| > 1$.

When $|\Gamma| > 1$, $(\bigcup_\Gamma \sim)$ is not necessarily transitive. This is the reason why it is not necessarily an equivalence relation, but it is still reflexive and symmetric.

Property 6.2

$$\widehat{M_{(\bigcap_\Gamma \sim)}} = \bigcap_{a \in \Gamma} \widehat{M_{\sim_a}} \text{ and } \forall a \in \Gamma : \widehat{M_{(\bigcup_\Gamma \sim)^+}} = \bigcup_{M' \in \widehat{M_{(\bigcup_\Gamma \sim)^+}}} \widehat{M'_{\sim_a}}.$$

According to this property, we have the following conclusion:

Property 6.3

$$\forall \mathbf{a} \in \Gamma: \quad \widehat{M_{(\cap_r \sim)}} \subseteq \widehat{M_{\sim_\mathbf{a}}} \subseteq \widehat{M_{(\cup_r \sim)^+}} \ .$$

Example 6.4 Figure 6.2d illustrates the equivalence relation $(\sim_{Alice} \cap \sim_{Bob})$, and it is coincident with the first conclusion of Property 6.2. It divides the reachable set into 9 equivalent classes (Fig. 6.2d and Table 6.1). From these 9 equivalent classes, we can observe that two states can be in the same class because a *super agent* who may be thought of as the combination of *Alice* and *Bob* owns the same knowledge at the two states. This point plays an important role in understanding the *distributed knowledge operator* in the next section.

Example 6.5 For $(\sim_{Alice} \cup \sim_{Bob})$, we know that

- $M_0(\sim_{Alice} \cup \sim_{Bob})M_2$ due to $M_0 \sim_{Alice} M_2$,
- $M_2(\sim_{Alice} \cup \sim_{Bob})M_3$ due to $M_2 \sim_{Bob} M_3$, but
- $(M_0, M_3) \notin (\sim_{Alice} \cup \sim_{Bob})$.

Therefore, $(\sim_{Alice} \cup \sim_{Bob})$ is not an equivalence relation. Based on this relation, we can compute its transitive closure $(\sim_{Alice} \cup \sim_{Bob})^+$ that exactly contains all the pairs of $M_0, M_1, \ldots,$ and M_{14}. The example is coincident with the second conclusion of Property 6.2, which helps to understand the *common knowledge operator* in the next section.

6.2 Computation Tree Logic of Knowledge

This section first recalls *computation tree logic of knowledge* (CTLK) including its syntax and semantics, and then introduces the computation methods of satisfiable state sets w.r.t. epistemic operators based on a reachability graph with equivalence relations. Finally, we introduce how to use the ROBDD technique to encode and compute those equivalence relations, and apply it to the verification of CTLK formulae.

6.2.1 Syntax and Semantics of CTLK

CTLK includes the operators of CTL as well as 4 epistemic operators. Here we describe the syntax and semantics directly according to the notations and notions of knowledge-oriented Petri nets [5]. Generally, this syntax and semantics were defined on the basis of other concurrent game structure or Kripke structure [9, 10], and readers may read these literatures for more details.

6.2.1.1 Syntax and Semantics of CTLK

Definition 6.2 (*Syntax of CTLK*) CTLK formulae can be defined according to the following grammar:

$$\phi ::= \textbf{true} \mid \textbf{false} \mid \textbf{deadlock} \mid p \mid \neg\phi \mid \phi_1 \vee \phi_2 \mid \phi_1 \wedge \phi_2 \mid \phi_1 \Rightarrow \phi_2 \mid$$

$$\textbf{EX } \phi \mid \textbf{EF } \phi \mid \textbf{EG } \phi \mid \textbf{E}(\phi_1 \textbf{ U } \phi_2) \mid \textbf{AX } \phi \mid \textbf{AF } \phi \mid \textbf{AG } \phi \mid \textbf{A}(\phi_1 \textbf{ U } \phi_2) \mid$$

$$\mathcal{K}_\textbf{a}\phi \mid \mathcal{E}_\Gamma\phi \mid \mathcal{D}_\Gamma\phi \mid \mathcal{C}_\Gamma\phi$$

where $\textbf{a} \in \mathcal{A}$, $\Gamma \subseteq \mathcal{A}$, and $p \in P_K \cup P_U$.

All operators except the last four epistemic operators have been explained in Chap. 5 and thus are not repeated here.

To explain the last four, a reachability graph with equivalence relations

$$RGER(N^\textbf{e}, M_0, \sim) = (R(N^\textbf{e}, M_0), E, \sim_{\textbf{a}_1}, \ldots, \sim_{\textbf{a}_k})$$

should be thought of as a labelled transition system.

$R(N^\textbf{e}, M_0)$ is the state set of this labelled transition system, M_0 is the initial state, E is the transition relation, all places form the label set (atomic propositions), and the labels associated with a marking (state) are those places that are marked at this marking. These equivalence relations are exactly used to explain the semantics of epistemic operators.

Definition 6.3 (*Semantics of Epistemic Operators*)

- $M \models \mathcal{K}_\textbf{a}\phi$ if and only if for each $M' \in R(N^\textbf{e}, M_0)$, if $M' \sim_\textbf{a} M$ then $M' \models \phi$,
- $M \models \mathcal{E}_\Gamma\phi$ if and only if for each $M' \in R(N^\textbf{e}, M_0)$, if $M'(\bigcup_\Gamma \sim)M$ then $M' \models \phi$,
- $M \models \mathcal{D}_\Gamma\phi$ if and only if for each $M' \in R(N^\textbf{e}, M_0)$, if $M'(\bigcap_\Gamma \sim)M$ then $M' \models \phi$,
- $M \models \mathcal{C}_\Gamma\phi$ if and only if for each $M' \in R(N^\textbf{e}, M_0)$, if $M'(\bigcup_\Gamma \sim)^+ M$ then $M' \models \phi$.

6.2.1.2 Explanation of $\mathcal{K}_\textbf{a}$

$M \models \mathcal{K}_\textbf{a}\phi$ means that the agent **a** knows the knowledge ϕ at the state M as well as all the other states equivalent to M.

6.2.1.3 Explanation of $\mathcal{E}_\Gamma$

$M \models \mathcal{E}_\Gamma \phi$ means that all agents in Γ know the knowledge ϕ at the state M as well as all the other states equivalent to M, i.e.,

$$\forall a \in \Gamma : M \models \mathcal{K}_a \phi.$$

Therefore, we have

$$M \models \mathcal{E}_\Gamma \phi = \bigwedge_{a \in \Gamma} (M \models \mathcal{K}_a \phi) = M \models \left(\bigwedge_{a \in \Gamma} \mathcal{K}_a \phi \right).$$

In other words, the operator $\mathcal{E}_\Gamma$ can be defined by $\mathcal{K}_a$ as follows:

$$\mathcal{E}_\Gamma \phi \triangleq \bigwedge_{a \in \Gamma} \mathcal{K}_a \phi.$$

Additionally, $\mathcal{K}_a$ is a special case of $\mathcal{E}_\Gamma$ where Γ has only one agent:

$$\mathcal{K}_a \phi \triangleq \mathcal{E}_{\{a\}} \phi.$$

In a word, $\mathcal{E}_\Gamma$ and $\mathcal{K}_a$ may be translated into each other.

6.2.1.4 Explanation of $\mathcal{D}_\Gamma$

$M \models \mathcal{D}_\Gamma \phi$ means that ϕ is a piece of distributed knowledge at M for all agents in Γ.

A piece of *distributed knowledge* at M is to say every agent in Γ knows a part of ϕ, but the *super agent* $\mathbf{a}_\Gamma$ combining all agents in Γ knows the whole of ϕ at M, i.e.,

$$\forall M' \in R(N^\mathfrak{k}, M_0): M' \sim_{\mathbf{a}_\Gamma} M \Rightarrow M' \models \phi,$$

where $\sim_{\mathbf{a}_\Gamma} \triangleq (\bigcap_\Gamma \sim)$. In other words,

$$M \models \mathcal{D}_\Gamma \phi = M \models \mathcal{K}_{\mathbf{a}_\Gamma} \phi.$$

Therefore, the operator $\mathcal{D}_\Gamma$ can also be defined by $\mathcal{K}_{\mathbf{a}_\Gamma}$, i.e.,

$$\mathcal{D}_\Gamma \phi \triangleq \mathcal{K}_{\mathbf{a}_\Gamma} \phi.$$

6.2.1.5 Explanation of C_Γ

> $M \models C_\Gamma \phi$ means that for all agents in Γ, ϕ is a piece of *common knowledge* at M.

In other words, when $M \models C_\Gamma \phi$, we have $M \models \mathcal{K}_\mathbf{a}\phi$ for each agent $\mathbf{a} \in \Gamma$, due to

$$\widehat{M_{\sim_\mathbf{a}}} \subseteq \widehat{M_{(\bigcup_\Gamma \sim)^+}}.$$

> Another important property caused by $M \models C_\Gamma \phi$ is the arbitrary transmission of the related knowledge among the agents in Γ.

Given an agent $\mathbf{a} \in \Gamma$, because

$$\widehat{M_{(\bigcup_\Gamma \sim)^+}} = \bigcup_{X \in \widehat{M_{(\bigcup_\Gamma \sim)^+}}} \widehat{X_{\sim_\mathbf{a}}},$$

we have that

$$\forall X \in \widehat{M_{(\bigcup_\Gamma \sim)^+}} : \ X \models \mathcal{K}_\mathbf{a}\phi.$$

In other words, if $\mathcal{K}_\mathbf{a}\phi$ is viewed as a piece of knowledge, then we have

$$\forall X \in \widehat{M_{(\bigcup_\Gamma \sim)^+}} : \ X \models C_\Gamma(\mathcal{K}_\mathbf{a}\phi).$$

Similarly, given another agent $\mathbf{b} \in \Gamma$, because

$$\widehat{M_{(\bigcup_\Gamma \sim)^+}} = \bigcup_{Y \in \widehat{M_{(\bigcup_\Gamma \sim)^+}}} \widehat{Y_{\sim_\mathbf{b}}},$$

we have that

$$\forall Y \in \widehat{M_{(\bigcup_\Gamma \sim)^+}} : \ Y \models \mathcal{K}_\mathbf{b}(\mathcal{K}_\mathbf{a}\phi).$$

Continually, we have that

$$\forall X \in \widehat{M_{(\bigcup_\Gamma \sim)^+}} : \ X \models \mathcal{K}_\mathbf{a}(\mathcal{K}_\mathbf{b}(\mathcal{K}_\mathbf{a}\phi)),$$

and so on. Therefore, this common knowledge operator can represent these pieces of knowledge can be arbitrarily transmitted among Γ. Notice that a piece knowledge of an agent can be arbitrarily transmitted in this agent itself, i.e.,

$$M \models \mathcal{K}_a \phi \Rightarrow M \models \mathcal{K}_a(\mathcal{K}_a \phi) \Rightarrow M \models \mathcal{K}_a(\mathcal{K}_a(\mathcal{K}_a \phi)) \Rightarrow \cdots.$$

In other words,

$$M \models \mathcal{K}_a \phi = M \models C_{\{a\}} \phi.$$

Example 6.6   For the key transport protocol, the CTLK formula

$$\mathbf{AG}(p_{1,7} \Rightarrow \mathcal{K}_{Alice}(\mathcal{K}_{Bob} p_{2,11})) \wedge \mathbf{AG}(p_{2,7} \Rightarrow \mathcal{K}_{Bob}(\mathcal{K}_{Alice} p_{1,8}))$$

is to represent that after the protocol is executed, Alice knows that Bob has got the password and Bob also knows that Alice has got the password. In what follows, we show it is valid in this protocol.

We see that the markings $M_0 - M_{13}$ all satisfy the sub-formula

$$p_{1,7} \Rightarrow \mathcal{K}_{Alice}(\mathcal{K}_{Bob} p_{2,11})$$

since the pre-condition $p_{1,7}$ does not hold at them. Only the marking M_{14} satisfies $p_{1,7}$. From Fig. 6.2c and Table 6.1, we see that for the agent *Bob*, the equivalent class including M_{14} is

$$\{M_{12}, \; M_{13}, \; M_{14}\},$$

and they all satisfy $p_{2,11}$. Hence, we have

$$M_{14} \models \mathcal{K}_{Bob} p_{2,11}.$$

For the agent *Alice*, the equivalent class including M_{14} is itself. Hence, we have

$$M_{14} \models \mathcal{K}_{Alice}(\mathcal{K}_{Bob} p_{2,11}).$$

Hence, all markings satisfy the sub-formula

$$p_{1,7} \Rightarrow \mathcal{K}_{Alice}(\mathcal{K}_{Bob} p_{2,11}),$$

and thus satisfy

$$\mathbf{AG}(p_{1,7} \Rightarrow \mathcal{K}_{Alice}(\mathcal{K}_{Bob} p_{2,11})).$$

Similarly, it can be verified that all markings also satisfy

$$\mathbf{AG}(p_{2,7} \Rightarrow \mathcal{K}_{Bob}(\mathcal{K}_{Alice} p_{1,8})).$$

Hence, we have

$$M_0 \models \mathbf{AG}(p_{1,7} \Rightarrow \mathcal{K}_{Alice}(\mathcal{K}_{Bob} p_{2,11})) \wedge \mathbf{AG}(p_{2,7} \Rightarrow \mathcal{K}_{Bob}(\mathcal{K}_{Alice} p_{1,8})),$$

i.e., this formula is valid in this system.

Example 6.7 Can the knowledge in the above example be transmitted continually? For instance, can Alice know that Bob has known that she has got the password after the protocol terminates? This instance can be represented by the following formula:

$$\mathbf{AG}(p_{2,7} \Rightarrow \mathcal{K}_{Alice}(\mathcal{K}_{Bob}(\mathcal{K}_{Alice}\, p_{1,8}))),$$

while the above common knowledge can be represented by the following one:

$$\mathbf{AG}(p_{2,7} \Rightarrow C_{\{Alice,Bob\}}\, p_{1,8}).$$

It is easy to check that the equivalence relation $(\sim_{Alice} \cup \sim_{Bob})^+$ leads to only one equivalent class, i.e., the whole reachable marking set itself. Obviously, some markings in this equivalent class (e.g. M_0 and M_2) do not satisfy $p_{1,8}$, and thus the above formula of common knowledge does not hold. Similarly, we can find that the first formula is invalid either, i.e., Alice does not know that Bob has known she has got the password.

6.2.2   Verifying CTLK Based on Reachability Graph with Equivalence Relations

This section introduces the computation of the satisfiable state set of a formula with different epistemic operators based on their semantics, that can be called in Algorithm 5.1 in Chap. 5. For more details, readers may read [1–3].

6.2.2.1   Computing the Satisfiable Marking Set of $\mathcal{K}_a \phi$

The computation of the satisfiable set of $\mathcal{K}_a \phi$ is to look for those equivalent classes in which every marking satisfies ϕ, as shown in Algorithm 6.1.

An easier procedure of computing $\mathbf{SAT}_{\mathcal{K}}(a, \phi)$ is to first compute all the markings satisfying $\neg \phi$ and then to compute their equivalent ones w.r.t. $\sim_a$. After removing all these computed markings out of the reachable marking set $R(N^{\varepsilon}, M_0)$, the remainders form $\mathbf{SAT}_{\mathcal{K}}(a, \phi)$. This procedure is illustrated at the bottom of Algorithm 6.1, and this idea can be used in the the following three ones.

Algorithm 6.1: SAT$_\mathcal{K}$(a, ϕ): Computing the Satisfiable Marking Set of $\mathcal{K}_a(\phi)$
Based on Reachability Graph with Equivalence Relations.

Input: A reachability graph with equivalence relations $(R(N^\mathfrak{e}, M_0), E, \sim)$, an agent **a**, and
       a formula ϕ;
Output: The satisfiable set **SAT$_\mathcal{K}$(a, ϕ)**;
1   $X \leftarrow$ **SAT**(ϕ);
2   $Y \leftarrow \emptyset$;
3   **for** *each equivalent class Z of* $\sim_a$ **do**
4   **if** $Z \subseteq X$ **then**
5         $\lfloor$ $Y \leftarrow Y \cup Z$;

6   **return** Y;

/* The following is an alternative procedure that is of a better operability than the above:

$X \leftarrow$ **SAT**$(\neg\phi)$;
$Y \leftarrow \{M \in R(N^\mathfrak{e}, M_0) \mid \exists M' \in X : M \sim_a M'\}$;
return $R(N^\mathfrak{e}, M_0) \setminus Y$;

*/

6.2.2.2   Computing the Satisfiable Marking Set of $\mathcal{E}_\Gamma \phi$

The computation of the satisfiable set of $\mathcal{E}_\Gamma \phi$ is to first compute the satisfiable set
of $\mathcal{K}_a\phi$ for each $a \in \Gamma$, and then the intersection of these sets is what we want, as
shown in Algorithm 6.2.

Similarly, a procedure that is more easily operated and understood is to first
compute the satisfiable marking set X of $\neg\phi$, and then for each agent $a \in \Gamma$ to
look for all possible markings equivalent to those in X. Then, the remainders form
SAT$_\mathcal{E}(\Gamma, \phi)$. This procedure is illustrated at the bottom of Algorithm 6.2.

6.2.2.3   Computing the Satisfiable Marking Set of $\mathcal{D}_\Gamma \phi$

For computing **SAT$_\mathcal{D}(\Gamma, \phi)$**, we first compute those markings satisfying $\neg\phi$. According to the semantics of $\mathcal{D}_\Gamma$, we have that for a given marking M' satisfying $\neg\phi$, if a
marking M is equivalent to M' for each agent $a \in \Gamma$, then M is not in **SAT$_\mathcal{D}(\Gamma, \phi)$**,
or it is what we want. Based on this idea, Algorithm 6.3 is designed to compute the
satisfiable set of such a formula.

6.2.2.4   Computing the Satisfiable Marking Set of $C_\Gamma \phi$

In view of the computation of **SAT$_\mathcal{E}(\Gamma, \phi)$**, it is easy to compute **SAT$_C(\Gamma, \phi)$** since
the semantics of $\mathcal{E}_\Gamma$ and C_Γ are based on

Algorithm 6.2: SAT$_\mathcal{E}$(Γ, ϕ): Computing the Satisfiable Marking Set of $\mathcal{E}_\Gamma\phi$ Based on Reachability Graph with Equivalence Relations.

Input: A reachability graph with equivalence relations $(R(N^\ell, M_0), E, \sim)$, a group Γ of
        agents, and a formula ϕ;
Output: The satisfiable set **SAT**$_\mathcal{E}$(Γ, ϕ);
1 $X \leftarrow$ **SAT**(ϕ);
2 $Y \leftarrow X$;
3 **for** *each agent* $\mathbf{a} \in \Gamma$ **do**
4   $\quad W \leftarrow \emptyset$;
5   $\quad$ **for** *each equivalent class Z of* $\sim_\mathbf{a}$ **do**
6   $\quad\quad$ **if** $Z \subseteq X$ **then**
7   $\quad\quad\quad W \leftarrow W \cup Z$;
8   $\quad Y \leftarrow Y \cap W$;
9 **return** Y;

/* The following is an alternative procedure that is of a better operability than the above:

$X \leftarrow$ **SAT**($\neg\phi$);
$Y \leftarrow \{M \in R(N^\ell, M_0) \mid \exists \mathbf{a} \in \Gamma, \exists M' \in X : M \sim_\mathbf{a} M'\}$;
return $R(N^\ell, M_0) \setminus Y$;

*/

Algorithm 6.3: SAT$_\mathcal{D}$(Γ, ϕ): Computing the Satisfiable Set of $\mathcal{D}_\Gamma\phi$ Based on Reachability Graph with Equivalence Relations.

Input: A reachability graph with equivalence relations $(R(N^\ell, M_0), E, \sim)$, a group Γ of
        agents, and a formula ϕ;
Output: The satisfiable set **SAT**$_\mathcal{D}$(Γ, ϕ);
1 $X \leftarrow$ **SAT**($\neg\phi$);
2 $Y \leftarrow \{M \in R(N^\ell, M_0) \mid \exists M' \in X, \forall \mathbf{a} \in \Gamma : M \sim_\mathbf{a} M'\}$;
3 **return** $R(N^\ell, M_0) \setminus Y$;

$$\left(\bigcup_\Gamma \sim\right) \text{ and } \left(\bigcup_\Gamma \sim\right)^+,$$

respectively.

In other words, we can repeatedly call the procedure of **SAT**$_\mathcal{E}$(Γ, ϕ) until the generated set is not changed any more. Algorithm 6.4 shows this procedure.

Question

Why do the computations of the satisfiable sets of $\mathcal{D}_\Gamma\phi$ and $\mathcal{C}_\Gamma\phi$ become complex if we do not use the ideas of first computing the satisfiable set of $\neg\phi$? Try to construct

---

Algorithm 6.4: $\text{SAT}_C(\Gamma, \phi)$: Computing the Satisfiable Marking Set of $C_\Gamma \phi$ Based on Reachability Graph with Equivalence Relations.

Input: A reachability graph with equivalence relations $(R(N^\mathfrak{e}, M_0), E, \sim)$, a group Γ of agents, and a formula ϕ;

Output: The satisfiable set $\text{SAT}_C(\Gamma, \phi)$;

1   $X \leftarrow \text{SAT}(\neg\phi)$;
2   $Y \leftarrow R(N^\mathfrak{e}, M_0)$;
3   **while** $X \neq Y$ **do**
4 | $Y \leftarrow X$;
5 | $X \leftarrow \{M \in R(N^\mathfrak{e}, M_0) \mid \exists M' \in Y, \exists \mathbf{a} \in \Gamma : M \sim_\mathbf{a} M'\}$;
6   **return** $R(N^\mathfrak{e}, M_0) \setminus Y$;

Algorithm 6.5: $\text{EQ}(\mathbb{M}, f, \mathbf{a})$: Generating the Equivalent Reachable Markings of a Given Group of Reachable Markings f w.r.t. the Agent $\mathbf{a}$.

Input: All reachable markings $\mathbb{M}$, a group of reachable markings f, and an agent $\mathbf{a}$;

Output: The reachable markings equivalent to the group of reachable markings f w.r.t. $\mathbf{a}$;

1   $Eq \leftarrow f$;
2   **for** *each* $p \in (P_K \cup P_U) \setminus P_\mathbf{a}$ **do**
3 | $Eq \leftarrow Eq \vee Eq[p \rightsquigarrow \neg p]$;
4   **return** $Eq \wedge \mathbb{M}$;

---

some formulae to illustrate the executions of these four algorithms, try to analyse the complexity of these algorithms, and try to write a related program.

6.2.3 Verifying CTLK Based on ROBDD

From Algorithms 6.1–6.4, we can see that even when no equivalence relation is stored in advance, those required equivalent markings can still be computed in the procedure of computing a satisfiable set. Therefore, it becomes possible to utilise the ROBDD technique to verify those formulae w.r.t. epistemic operators, especially when only reachable markings are encoded and stored but neither transition relations nor equivalence relations are stored. Next, we will introduce such a procedure, and readers may find more details in [4–6].

From Algorithms 6.1–6.4, we can see that the key is to compute those equivalent markings of a given set of ones (w.r.t. a given agent). Algorithm 6.5 shows the procedure of computing all the equivalent reachable markings for a given set of markings and a given agent.

It is very easy to carry out this task by using the ROBDD technique. First, similar to the algorithms in Chap. 5, $\mathbb{M}$ stands for the ROBDD representing all reachable markings of a knowledge-oriented Petri net, and f stands for the ROBDD representing a given group of markings. Certainly, these ROBDDs can be thought of as Boolean functions. To compute those equivalent markings of f w.r.t. an agent $\mathbf{a}$, we may compute all possible markings equivalent to f, i.e., computing a Boolean function Eq in which all variables except those ones in P_a can be assigned any one of **true** and **false** (Lines 2–3 in Algorithm 6.5). Since some markings in the set represented by Eq are not reachable, the operation $\mathbb{M} \wedge Eq$ yields what we want (Line 4 in Algorithm 6.5).

Example 6.8 We use the agent Bob and markings $\{M_1, M_{13}\}$ in Fig. 6.2c to illustrate the computing of their equivalent markings. Certainly, from Fig. 6.2c we can see that their equivalent markings should be $\{M_0, M_1, M_4, M_{12}, M_{13}, M_{14}\}$. First, $\{M_1, M_{13}\}$ is represented by a Boolean function as follows:

$$- \Big(\neg p_{1,1} \wedge p_{1,2} \wedge \neg p_{1,3} \wedge \neg p_{1,4} \wedge \neg p_{1,5} \wedge \neg p_{1,6} \wedge \neg p_{1,7} \wedge p_{1,8} \wedge \neg p_{1,9} \wedge \neg p_{1,10}$$
$$\wedge \neg p_{1,11} \wedge$$

$$p_{2,1} \wedge \neg p_{2,2} \wedge \neg p_{2,3} \wedge \neg p_{2,4} \wedge \neg p_{2,5} \wedge \neg p_{2,6} \wedge \neg p_{2,7} \wedge \neg p_{2,8} \wedge \neg p_{2,9} \wedge \neg p_{2,10} \wedge \neg p_{2,11} \wedge$$

$$\neg p_1 \wedge \neg p_2 \wedge \neg p_3 \wedge \neg p_4 \Big) \vee$$

$$\Big(\neg p_{1,1} \wedge \neg p_{1,2} \wedge \neg p_{1,3} \wedge \neg p_{1,4} \wedge \neg p_{1,5} \wedge p_{1,6} \wedge \neg p_{1,7} \wedge p_{1,8} \wedge p_{1,9} \wedge p_{1,10} \wedge \neg p_{1,11} \wedge$$

$$\neg p_{2,1} \wedge \neg p_{2,2} \wedge \neg p_{2,3} \wedge \neg p_{2,4} \wedge \neg p_{2,5} \wedge \neg p_{2,6} \wedge p_{2,7} \wedge p_{2,8} \wedge p_{2,9} \wedge p_{2,10} \wedge p_{2,11} \wedge$$

$$\neg p_1 \wedge \neg p_2 \wedge \neg p_3 \wedge p_4 \Big).$$

Since $P_{Bob} = \{p_{2,8}, p_{2,9}, p_{2,10}, p_{2,11}\}$, we execute the **for** loop in Algorithm 6.5 to remove those variables that are not in P_{Bob} and then obtain the following Boolean function:

$$- \Big(\neg p_{2,8} \wedge \neg p_{2,9} \wedge \neg p_{2,10} \wedge \neg p_{2,11} \Big) \vee \Big(p_{2,8} \wedge p_{2,9} \wedge p_{2,10} \wedge p_{2,11} \Big).$$

The conjunction of this Boolean function and $\mathbb{M}$ exactly leads to the Boolean function representing $\{M_0, M_1, M_4, M_{12}, M_{13}, M_{14}\}$, and Table 6.1 can help readers understand this result.

Based on Algorithms 6.1–6.5 as well as the ROBDD technique, we can provide the procedure of computing the satisfiable marking set of a CTLK formula with epistemic

---

Algorithm 6.6: SAT$(\mathbb{M}, \phi)$: Computing the Satisfiable Marking Set of CTLK Formula ϕ Based on the ROBDD Technique.

Input: The ROBDD $\mathbb{M}$ encoding all reachable markings of a given knowledge-oriented Petri net and a CTLK formula ϕ;

Output: The satisfiable set **SAT**$(\mathbb{M}, \phi)$;

1 **if** $\phi = $ **true then**
2 $\quad$ | $\quad$ return $\mathbb{M}$;

3 **if** $\phi = p$ **then**
4 $\quad$ | $\quad$ return $p \wedge \mathbb{M}$;

5 **if** $\phi = \neg\phi_1$ **then**
6 $\quad$ | $\quad$ return $\mathbb{M} - $ **SAT**$(\mathbb{M}, \phi_1)$;

7 **if** $\phi = \phi_1 \wedge \phi_2$ **then**
8 $\quad$ | $\quad$ return **SAT**$(\mathbb{M}, \phi_1) \wedge$ **SAT**$(\mathbb{M}, \phi_2)$;

9 **if** $\phi = $ **EX** ϕ_1 **then**
10 $\quad$ | $\quad$ return **PRE**$(\mathbb{M}, $ **SAT**$(\mathbb{M}, \phi_1))$;

11 **if** $\phi = $ **EG** ϕ_1 **then**
12 $\quad$ | $\quad$ $X \leftarrow $ **SAT**$(\mathbb{M}, \phi_1)$; $Z \leftarrow X \wedge (\mathbb{M} - ((\bigvee_{t \in T} \bigwedge_{p \in {}^\bullet t} p) \wedge \mathbb{M}))$;
13 $\quad$ | $\quad$ $Y \leftarrow ($ **PRE**$(\mathbb{M}, X) \wedge X) \vee Z$;
14 $\quad$ | $\quad$ **while** $X \neq Y$ **do**
15 $\quad$ | $\quad$ $\quad$ | $\quad$ $X \leftarrow Y$; $Y \leftarrow ($ **PRE**$(\mathbb{M}, X) \wedge X) \vee Z$;
16 $\quad$ | $\quad$ return Y;

17 **if** $\phi = $ **E**$(\phi_1$ **U** $\phi_2)$ **then**
18 $\quad$ | $\quad$ $X \leftarrow $ **SAT**$(\mathbb{M}, \phi_1)$; $Y \leftarrow $ **SAT**$(\mathbb{M}, \phi_2)$;
19 $\quad$ | $\quad$ **while** $X \neq $ *false* **do**
20 $\quad$ | $\quad$ $\quad$ | $\quad$ $Z \leftarrow X \wedge $ **PRE**$(\mathbb{M}, Y)$; $X \leftarrow X - Z$; $Y \leftarrow Y \vee Z$;
21 $\quad$ | $\quad$ return Y;

22 **if** $\phi = \mathcal{K}_\mathbf{a}\phi_1$ **then**
23 $\quad$ | $\quad$ return $\mathbb{M} - $ **EQ**$(\mathbb{M}, $ **SAT**$(\mathbb{M}, \neg\phi_1), \mathbf{a})$;

24 **if** $\phi = \mathcal{E}_\Gamma\phi_1$ **then**
25 $\quad$ | $\quad$ $X \leftarrow $ **SAT**$(\neg\phi_1)$; $Y \leftarrow $ **false**;
26 $\quad$ | $\quad$ **for** *each* $\mathbf{a} \in \Gamma$ **do**
27 $\quad$ | $\quad$ $\quad$ | $\quad$ $Y \leftarrow Y \vee $ **EQ**$(\mathbb{M}, X, \mathbf{a})$;
28 $\quad$ | $\quad$ return $\mathbb{M} - Y$;

29 **if** $\phi = \mathcal{D}_\Gamma\phi_1$ **then**
30 $\quad$ | $\quad$ $X \leftarrow $ **SAT**$(\neg\phi_1)$; $Y \leftarrow \mathbb{M}$;
31 $\quad$ | $\quad$ **for** *each* $\mathbf{a} \in \Gamma$ **do**
32 $\quad$ | $\quad$ $\quad$ | $\quad$ $Y \leftarrow Y \wedge $ **EQ**$(\mathbb{M}, X, \mathbf{a})$;
33 $\quad$ | $\quad$ return $\mathbb{M} - Y$;

34 **if** $\phi = C_\Gamma\phi_1$ **then**
35 $\quad$ | $\quad$ $X \leftarrow \mathbb{M}$; $Y \leftarrow $ **SAT**$(\neg\phi_1)$;
36 $\quad$ | $\quad$ **while** $X \neq Y$ **do**
37 $\quad$ | $\quad$ $\quad$ | $\quad$ $X \leftarrow Y$; $Y \leftarrow $ **false**;
38 $\quad$ | $\quad$ $\quad$ | $\quad$ **for** *each* $\mathbf{a} \in \Gamma$ **do**
39 $\quad$ | $\quad$ $\quad$ | $\quad$ $\quad$ | $\quad$ $Y \leftarrow Y \vee $ **EQ**$(\mathbb{M}, X, \mathbf{a})$;
40 $\quad$ | $\quad$ return $\mathbb{M} - Y$;

---

operators, as shown in Lines 22–40 in Algorithm 6.6. Notice that Lines 1–21 in Algorithm 6.6 form exactly Algorithm 5.11. Readers may verify these procedures through writing a program (by calling the package of ROBDD) and constructing some net systems and some formulae.

6.3   Applications

This section involves two examples to illustrate the usefulness of knowledge-oriented Petri nets and CTLK.

6.3.1   Key Transport Protocol with Attacking

Section 6.1 introduces this protocol. Here we show that it is insecure when the messages are transported in an insecure channel. We assume that the messages can be intercepted by an attacker. The knowledge-oriented Petri net in Fig. 6.3 models this protocol with an attacker.

This attacker (the agent *Attacker*) first intercepts the request of Alice ($t_{3,2}$), and then sends her/his public key to Alice ($t_{3,3}$). After Alice uses this public key to encrypt her password and sends it to Bob, the attacker intercepts it again ($t_{3,4}$). Then the attacker uses her/his private key to decrypt it and thus obtains this password ($t_{3,5}$). Finally, the attacker sends Alice an acknowledgement ($t_{3,6}$).

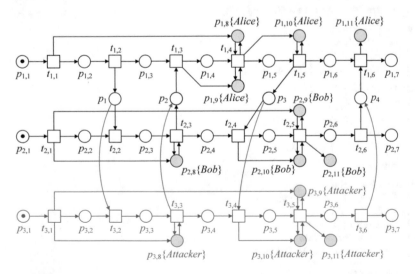

Fig. 6.3 The knowledge-oriented Petri net modelling the key transport protocol with public-key encryption, considering an attacker

This protocol is insecure since the attacker can obtain Alice's password but Bob and Alice cannot know it. This property can be represented by the following formula:

$$\mathbf{EF}(p_{3,11} \wedge \neg \mathcal{K}_{\{Alice\}} p_{3,11} \wedge \neg \mathcal{K}_{\{Bob\}} p_{3,11})$$

which is valid in this system [6].

6.3.2 Dining Cryptographers Protocol

6.3.2.1 Protocol Description and Model

The dining cryptographers protocol is an abstract example of anonymity protocols [11], aiming at protecting privacy during an exchange. Three cryptographers share a meal around a table and either one of them or their employer pays for the meal. The designed protocol requires [10, 12, 13] that

- if their employer paid then they all know it, and
- if one of these cryptographers paid then they know that one of them paid but do not know who paid.

There is a coin between any two cryptographers, and every coin can be randomly tossed. After a coin between two cryptographers is tossed, the result (i.e., head or tail) can only be seen by the two cryptographers rather than the third one. After a cryptographer sees the results of the two tossed coins associated with her/him, s/he makes an announcement, i.e., saying "same" or "different". The protocol requires that a cryptographer tells a lie if s/he pays money, or a truth.

The KPN in Fig. 6.4 models this protocol about the case with three cryptographers. Here, the agent a_0 stands for the employer, and the agent a_j stands for the jth cryptographer where $j \in \mathbb{N}_3^+$. Firing the transition $t_{j,0}$ means that the agent a_j paid where $j \in \mathbb{N}_3$. Since only one of the four agents will pay, the four transitions $t_{j,0}$ ($j \in \mathbb{N}_3$) share the common place p_0 as their inputs. The inputs of the transition $t_{1,2}$ include the places $paid_1$, $head_1$, and $head_2$; hence, firing $t_{1,2}$ means that the agent a_1 paid and saw two heads; hence, this agent said "different", i.e., the output of $t_{1,2}$ is $different_1$. Similarly, readers can understand the meanings of other transitions.

6.3.2.2 Requirement Representation and Verfification

The two design requirements mentioned above can be represented by the following two CTLK formulae:

- $\mathbf{AG}\left(\left(p_{1,3} \wedge p_{2,3} \wedge p_{3,3} \wedge paid_0 \right) \Rightarrow C_{\{a_1,a_2,a_3\}} paid_0 \right)$, and

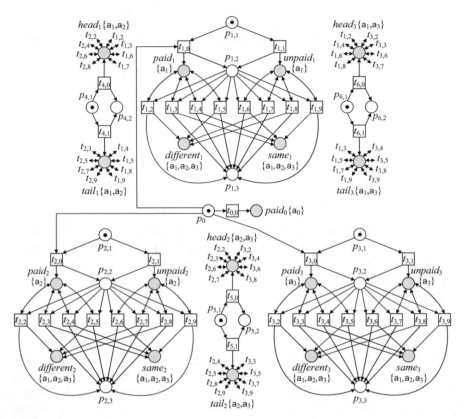

Fig. 6.4 The knowledge-oriented Petri net modelling the dining cryptographers protocol. Notice, for simplification, a self-loop is represented by an arc with arrowheads at both ends

$$- \mathbf{AG}\Bigg(\Big(p_{1,3} \wedge p_{2,3} \wedge p_{3,3} \wedge \neg paid_0 \wedge \neg paid_1 \Big) \Rightarrow$$

$$\Big(\mathcal{K}_{\mathbf{a}_1}\big(paid_2 \vee paid_3 \big) \wedge \neg \mathcal{K}_{\mathbf{a}_1} paid_2 \wedge \neg \mathcal{K}_{\mathbf{a}_1} paid_3 \big) \Bigg).$$

which are both valid in this protocol [5, 10, 12, 13]. The first formula represents that after the employer paid, every cryptographer knows it and knows that other cryptographers know it. The second one represents that after one of the second or third cryptographer paid, the first cryptographer knows it but does know which one of the two cryptographers paid. Due to the symmetry among the three cryptographers, we can write the related formulae for the second and third cryptographers, similar to the above second formula.

When we consider n ($n \geq 3$) cryptographers, the above two formulae can be extended to the following forms to represent these design requirements:

$$- \text{ } \textbf{AG}\left(\left(\bigwedge_{j=1}^{n} p_{j,3} \wedge paid_0\right) \Rightarrow C_{\mathcal{A}} paid_0\right),$$

$$- \text{ } \textbf{AG}\left(\left(\bigwedge_{j=1}^{n} p_{j,3} \wedge \neg paid_0 \wedge \neg paid_1\right) \Rightarrow \left(\mathcal{K}_{\mathbf{a}_1}\left(\bigvee_{j=2}^{n} paid_j\right) \wedge \bigwedge_{j=2}^{n} \neg\mathcal{K}_{\mathbf{a}_1} paid_j\right)\right),$$

where $\mathcal{A} = \{\mathbf{a}_1, \mathbf{a}_2, \ldots, \mathbf{a}_n\}$. When we use the ROBDD technique as well as a variable order computed by the method in [14], we can check the case of 1200 cryptographers in 14 hours over a PC [5]. This case has about 10^{1080} reachable markings, and the two checked formulae have more than 6000 atomic propositions and more than 3600 operators, which is amazing!

6.4   Summery and Further Reading

Knowledge-oriented Petri nets can both model the interaction process of multiple agents and represent the epistemic evolution of each agent. They not only have the intuitiveness for modelling these multi-agent systems or protocols, but also have advantages on analysing such a large-scale system since the reachable markings of this type of Petri nets are easily compressed and those equivalence relations are not required to be stored in advance (whereas some other modelling languages and analysis methods have to store these equivalence relations in advance [10, 12, 13]). This chapter briefly introduces the basic principle of checking CTLK on the basis of knowledge-oriented Petri nets.

Smart contracts play an important role in the block-chain technique, and there have been some methods to verify them [15, 16]. How to represent more complex knowledge as well as more privacy properties using Petri nets and temporal logics and how to apply them in the verification of smart contracts are interesting topics.

References

1. He, L.F., Liu, G.J.: Model checking CTLK based on knowledge-oriented Petri nets. In: *the 21st IEEE International Conference on High Performance Computing and Communications*, pp. 1139–1146 (2019)
2. He, L.F., Liu, G.J.: Petri nets based verification of epistemic logic and its application on protocols of privacy and security. In: *2020 IEEE World Congress on Services*, pp. 25–28 (2020)
3. He, L.F., Liu, G.J.: Verifying computation tree logic of knowledge via the similar reachability graph of knowledge-oriented Petri nets. In: *the 39th Chinese Control Conference*, pp. 5026–5031 (2020)
4. He, L.F., Liu, G.J.: Verifying computation tree logic of knowledge via knowledge-oriented Petri nets and ordered binary decision diagrams. *Computing and Informatics* 40: 1174–1196 (2021)

5. He, L.F., Liu, G.J., Zhou, M.C.: Petri-net based model checking for privacy-critical multiagent systems. *IEEE Transactions on Computational Social systems*, (2022)
6. He, L.F., Liu, G.J.: Petri net based symbolic model checking for computation tree logic of knowledge, arXiv:2012.10126v1, 1–23 (2020)
7. Paar, C., Pelzl, J.: *Understanding Cryptography: A Textbook for Students and Practitioners.* Springer Science and Business Media (2009)
8. Jongsma, C.: *Introduction to Discrete Mathematics via Logic and Proof.* Springer Nature (2019)
9. Alur, R., Henzinger, T.A., Kupferman, O.: Alternating-time temporal logic. *Journal of the ACM* 49: 672–713 (2002)
10. Lomuscio, A., Qu, H., Raimondi, F.: MCMAS: An open-source model checker for the verification of multi-agent systems. *International Journal on Software Tools for Technology Transfer* 19: 9–30 (2017)
11. Chaum, D.: The dining cryptographers problem: Unconditional sender and recipient untraceability. *Journal of Cryptology* 1: 65–75 (1988)
12. P. Gammie, P., van der Meyden, R.: MCK: Model checking the logic of knowledge, in: *the 16th International Conference on Computer Aided Verification*, pp. 479–483 (2004)
13. Su, K.L., Sattar, A., Luo, X.: Model checking temporal logics of knowledge via OBDDs, *the Computer Journal* 50: 403–420 (2007)
14. He, L.F., Liu, G.J.: Petri net based CTL model checking: using a new method to construct OBDD variable order. In: *the 15th International Symposium on Theoretical Aspects of Software Engineering*, pp. 159–166 (2021)
15. Liu, J., Liu, Z.: A survey on security verification of blockchain smart contracts. *IEEE Access* 7: 77894–77904 (2019)
16. Singh, A., Parizi, R.M., Zhang, Q., Choo, K.K.R., Dehghantanha, A.: Blockchain smart contracts formalization: Approaches and challenges to address vulnerabilities. *Computers and Security* 88, article no. 101654 (2020)

Chapter 7
Petri Nets with Insecure Places and Secure Bisimulation

7.1 Bisimulation and Weak Bisimulation

This section recalls the notions of bisimulation and weak bisimulation, and readers can find more details in [1, 2].

In the classic theory of formal languages and automata, two automata are thought of having an equivalent behaviour if they recognise the same language [3, 4]. Roughly speaking, two automata recognise the same language if and only if they have the same set of action sequences. Later, Milner found that this classic notion of language equivalence is not reasonable to explain some real applications. The famous example given by him is about two tea/coffee-vending machines as shown in Fig. 7.1 [2]: one is deterministic and another is nondeterministic. In other words, those classic automata models like Turing machines, register machines, and the lambda calculus are concerned with *computational behaviour* but not *interactional behaviour*. A basic action of an interactive system is to communicate across an interface with a handshake. He found that the determinism/nondeterminism of the internal actions of a system can affect the interacting behaviour while the notion of language equivalence does not consider the difference between determinism and nondeterminism. He thus proposed the notion of (weak) bisimulation equivalence [1, 2]. If two systems are (weakly) bisimilar, they recognise the same language, but not vice versa. The (weak) bisimulation equivalence takes account of a more detailed relationship of states than the language equivalence does.

An automaton $\mathbb{A} = (St, Act, \longrightarrow, s_0)$ is *deterministic* if it has no ε-transition and

$$(s \xrightarrow{a} s' \wedge s \xrightarrow{a} s'') \Rightarrow s' = s'',$$

or else it is *nondeterministic*, where $\varepsilon \in Act$ stands for *unobservable actions* and $a \in Act$ is an *observable action*.

Example 7.1 In Fig. 7.1, (a) shows a deterministic automaton and (b) a nondeterministic one in which there is no ε-transition. This deterministic machine means that a buyer can get a cup of tea if s/he wants it and puts in 2 pence, or s/he can get a cup of coffee if she wants it and puts in 4 pence. This nondeterministic one means that after a buyer put in 2 pence, the machine may be at the state s_3 at which s/he has to

Fig. 7.1 Two different automata modelling a tea/coffee-vending machine [2]. **a** deterministic and **b** nondeterministic

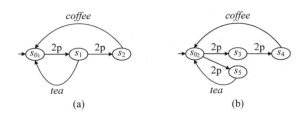

put in 2 more pence to get a cup of coffee (so that s/he cannot get tea if s/he wants it), or it may be at the state s_5 at which s/he only gets a cup of tea (so that s/he cannot get coffee if s/he wants it). Obviously, the two machines are not equivalent, but they are thought of being equivalent according to the language equivalence theory since they recognise the same language

$$(2\text{p} \cdot tea + 2\text{p} \cdot 2\text{p} \cdot coffee)^*.$$

In what follows, an automaton is non-deterministic in default unless some specific constraints are provided. If

$$s_1 \xrightarrow{\varepsilon} s_2 \xrightarrow{\varepsilon} \cdots \xrightarrow{\varepsilon} s_n,$$

then it is denoted as $s_1 \dashrightarrow s_n$. We assume that $\dashrightarrow$ is reflexive, i.e., $s \dashrightarrow s$. If

$$s_1 \dashrightarrow s_n \xrightarrow{a} s_{n+1},$$

then it is denoted as $s_1 \overset{a}{\dashrightarrow} s_{n+1}$.

Definition 7.1 (*Bisimulation of Two Automata*) Let $\mathbb{A}_1 = (St_1, Act, \longrightarrow_1, s_{0_1})$ and $\mathbb{A}_2 = (St_2, Act, \longrightarrow_2, s_{0_2})$ be two automata such that $St_1 \cap St_2 = \emptyset$ and $\varepsilon \notin Act$. A binary relation $\approx \subseteq (St_1 \times St_2) \cup (St_2 \times St_1)$ is a *bisimulation* if the following two conditions hold:

1. $\approx$ is symmetric, and
2. if $s_1 \approx s_2$ and $s_1 \longrightarrow_1 s_1'$, then there exists s_2' such that $s_2 \longrightarrow_2 s_2'$ and $s_1' \approx s_2'$.

If there is a bisimulation $\approx$ such that $s_{0_1} \approx s_{0_2}$, then $\mathbb{A}_1$ and $\mathbb{A}_2$ are called *bisimilar*, denoted as $\mathbb{A}_1 \approx \mathbb{A}_2$.

Example 7.2 The two automata in Fig. 7.1 are not bisimilar. If they were bisimilar, then there should be $s_{0_1} \approx s_{0_2}$, and thus $s_1 \approx s_3$. However, there is a *tea*-transition outgoing at s_1 in (a):

$$s_1 \xrightarrow{tea} s_{0_1},$$

but there is no any *tea*-transition outgoing at s_3 in (b). Therefore, the bisimulation equivalence can distinguish the two automata.

Definition 7.2 (*Weak Bisimulation of Two Automata*) Let $\mathbb{A}_1 = (St_1, Act, \longrightarrow_1, s_{0_1})$ and $\mathbb{A}_2 = (St_2, Act, \longrightarrow_2, s_{0_2})$ be two automata such that $St_1 \cap St_2 = \emptyset$. A binary relation $\approx \subseteq (St_1 \times St_2) \cup (St_2 \times St_1)$ is a *weak bisimulation* (or *observational equivalence*) if the following three conditions hold:

1. $\approx$ is symmetric,
2. if $s_1 \approx s_2$ and $s_1 \xrightarrow{\varepsilon}_1 s_1'$, then there exists s_2' such that $s_2 \dashrightarrow_2 s_2'$ and $s_1' \approx s_2'$,
3. if $s_1 \approx s_2$ and $s_1 \xrightarrow{a}_1 s_1'$, then there exists s_2' such that $s_2 \overset{a}{\dashrightarrow}_2 s_2'$ and $s_1' \approx s_2'$.

If there is a weak bisimulation $\approx$ such that $s_{0_1} \approx s_{0_2}$, then $\mathbb{A}_1$ and $\mathbb{A}_2$ are called *weakly bisimilar*, denoted as $\mathbb{A}_1 \approx \mathbb{A}_2$.

Weak bisimulation considers the automata with unobservable actions, and later we will see a related example. Obviously, bisimulation is a special case of weak bisimulatin, because the second condition does not occur when two automata have no unobservable actions.

> The (weak) bisimulation equivalence takes into account not only all possible action sequences but also all possible actions at each state.

Therefore, they are more accurate in distinguishing two automata than the language equivalence does.

Theorem 7.1 ([2]) *If two automata are equivalent according to the bisimulation equivalence, then they are also equivalent according to the language equivalence; but not vice versa.*

Another thing which is worthy noting is that bisimulation and weak bisimulation are defined over the states of an automaton in [1, 2]. Here they are defined over the states of two automata but the action sets of the two automata are identical. When a (weak) bisimulation is defined over an automaton, it is an equivalence relation. A bisimulation is called a *strong bisimulation* in [1, 2], and for simplicity we omit "strong" in this book.

In the next section, we will introduce a system of transferring money online. When this system takes two different security policies, it is easily found that one has a stronger ability to resist an attack than another. However, the systems caused by the two policies have the same behaviours whether we use the language equivalence or the bisimulation equivalence to evaluate their behaviours. We hope a new notion of behaviour equivalence can distinguish them, which is the motivation of this chapter.

? Question

Try to design an algorithm to decide if two automata are (weakly) bisimilar.

7.2   Petri Nets with Insecure Places and Secure Bisimulation

This section introduces the notion of Petri nets with insecure places and their reachability graphs, and then provides the definition of secure bisimulation. Readers can find more details in [5, 6].

7.2.1   Petri Nets with Insecure Places

Definition 7.3 (*Petri Net with Insecure Places*) A *Petri net with insecure places* $(N^{is}, M_0) \triangleq (P_S \cup P_I, T, F, M_0)$ is a Petri net where P_S is a finite set of *secure places*, P_I is a finite set of *insecure places*, and $P_S \cap P_I = \emptyset$.

A Petri net with insecure places is actually a Petri net in which places are divided into *secure* and *insecure*. But, this division does not destroy the enabling and firing rules of transitions, just like a knowledge-oriented Petri net. If a reachable marking marks an insecure place, it is an *insecure marking*.

All insecure markings of a Petri net with insecure places are denoted as $\mathbb{IM}$. Notice that all transitions of a Petri net with insecure places are observable here.

Next, we consider two versions of modelling the electronic funds transfer system of a bank. The two systems are sketched as shown in Fig. 7.2a, b. Here we only consider a very simple workflow including only the steps of *submitting a transferring-money request, receiving a verification message*, and *agreeing or disagreeing with this transfer*.

First, a user fills in the information of transferring money with a PC (the transition t_1) and sends the bank system a request (the transition t_2). Notice that the filled information includes money amount, the target account, and so on; and the submitted request is first packaged and encrypted in the PC and then is sent from the PC to the bank system located in the bank. Packaging and encrypting this request is modelled by the place $p_{1,3}$ in Fig. 7.2a or $p_{2,3}$ in Fig. 7.2b. After receiving the request (t_3), the bank system sends the user a verification message (the transition t_4). This message includes the source account, the target account, the transferred money amount, and a verification code. Notice that for simplicity, we do not consider the step of the bank system checking if the request is legal. Firing the transition t_5 means that the PC receives this message, and firing the transition t_6 means that this message is shown to the user. The places $p_{1,9}$ in Fig. 7.2a and $p_{2,9}$ in Fig. 7.2b means that this message has reached the PC but has not been shown to the user, and the places $p_{1,10}$ and $p_{2,10}$ means that it has been seen by the user. Firing the transition t_7 means that the user disagrees with this transfer and cancels it, while firing t_8 means that the user inputs the verification code and agrees with this transfer, according to the message s/he saw.

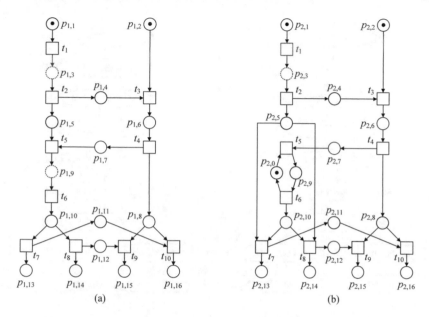

Fig. 7.2 Two Petri nets with insecure places modelling the electronic funds transfer system of a bank. **a** taking the policy that the operations of transferring money and the verification are conducted in the same device, but **b** in different devices. Notice that the dotted circle stands for insecure places

The two systems have the same function and workflow. They are equivalent according to the bisimulation equivalence since we easily construct a bisimulation based on their rechabliility graphs as shown in Fig. 7.3.

The distinction between them is that the system in Fig. 7.2a sends the verification message to the PC itself while the one in Fig. 7.2b sends it to another independent device (e.g. the user's cell phone). In other words, the former means that the devices of logging in the funds transfer system and receiving the verification message are the same but the latter means that the two devices are different.

At present, banks do not think that they are equivalent. They usually use the system in Fig. 7.2b rather than the one in (a) because they believe that the former is much securer than the latter.[1] For the system in Fig. 7.2a, a Trojan program can transfer the user's money from the account of this user into an account that is not the target one of this user.

For example, a Trojan program is implanted into the user's PC. The Trojan program can tamper with the data in the transferring-money request of the user during

---

[1] The security of a system is related to many factors including crypto-systems, rights of accessing data, and other supporting policies like the one of separating "verification" from "login" in Fig. 7.2b. The security in this paper is about those supporting policies rather than crypto-systems.

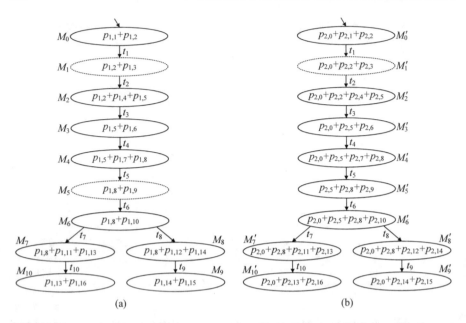

Fig. 7.3 The reachability graphs with insecure markings of the two Petri nets with insecure places in Fig. 7.2. Notice that dotted eclipses stand for insecure markings. They recognise the same language $\{t_1t_2t_3t_4t_5t_6(t_7t_{10} + t_8t_9)\}$, and they are bisimilar since there exists a bisimulation $\{(M_j, M_j'), (M_j', M_j) \mid j \in \mathbb{N}_{10}\}$ such that $M_0 \approx M_0'$

the period of packaging this request, and can also tamper with the verification message before it is shown to the user. While a legal request of transferring money from Account 1 to Account 2 is packaged, the Trojan changes it into the one of transferring money from Account 1 to Account 3. Thus, the tampered request is sent to the bank system. The bank system finds that Accounts 1 and 3 are both legal and then sends the user a verification message in which the bank tells the user that money will be transferred into Account 3 and provides the user a verification code. When the PC has received the verification message but not yet shown it to the user, the Trojan changes Account 3 in this verification message into Account 2. Finally, the user finds that all are correct and then inputs the verification code to agree with this transfer. Unfortunately, the money is actually transferred into Account 3 rather than Account 2. It is almost impossible for the system in Fig. 7.2b that a Trojan program is simultaneously implanted into both the PC and the cell phone of a user, and thus the above case does not occur. Notice that the places $p_{1,3}$, $p_{2,3}$, and $p_{1,9}$ are defined as *insecure* just because the information represented by them can be tampered with. Therefore, the two systems are not *equivalent*: they produce different results for the same input. The two different security policies lead to different interacting behaviours between the system and its users since they handle the external nondeterministic events through different measures. This is exactly the reason that the policy in Fig. 7.2b can enhance the security. Therefore, the bisimulation notion cannot directly be applied to this case.

The question is: how to use a formal method to declare the (in-)equivalence of two systems with the above different security policies? To this end, we propose secure bisimulation in what follows.

7.2.2 Secure Bisimulation

Definition 7.4 (*Secure Bisimulation of Two Petri Nets with Insecure Places*) Let $(N_1^{is}, M_{0_1}) = (P_{S_1} \cup P_{I_1}, T, F_1, M_{0_1})$ and $(N_2^{is}, M_{0_2}) = (P_{S_2} \cup P_{I_2}, T, F_2, M_{0_2})$ be two Petri nets with insecure places such that $(P_{S_1} \cup P_{I_1}) \cap (P_{S_2} \cup P_{I_2}) = \emptyset$, and let $\mathbb{IM}_1$ and $\mathbb{IM}_2$ be their insecure marking sets. A binary relation

$$|\widetilde{\approx}| \subseteq (R(N_1^{is}, M_{0_1}) \times R(N_2^{is}, M_{0_2})) \cup (R(N_2^{is}, M_{0_2}) \times R(N_1^{is}, M_{0_1}))$$

is a *secure bisimulation* if the following three conditions hold:

1. $|\widetilde{\approx}|$ is symmetric,
2. if $M_1 |\widetilde{\approx}| M_2$ and $M_1[t\rangle M_1'$, then there exists M_2' such that $M_2[t\rangle M_2'$ and $M_1' |\widetilde{\approx}| M_2'$,
3. if $M_1 |\widetilde{\approx}| M_2$, then M_1 and M_2 are either in $\mathbb{IM}_1 \cup \mathbb{IM}_2$ or not in $\mathbb{IM}_1 \cup \mathbb{IM}_2$.

If there is a secure bisimulation $|\widetilde{\approx}|$ such that $M_{0_1} |\widetilde{\approx}| M_{0_2}$, then (N_1^{is}, M_{0_1}) and (N_2^{is}, M_{0_2}) are called *securely bisimilar*, denoted as $(N_1^{is}, M_{0_1}) |\widetilde{\approx}| (N_2^{is}, M_{0_2})$.

The third condition means that for two markings in a secure bisimulation, either they are both secure, or they are both insecure. Obviously, if a secure bisimulation is defined over two Petri nets without insecure place, then the secure bisimulation is a bisimulation. Based on secure bisimulation, we know that the two Petri nets with insecure places in Fig. 7.2 are not securely bisimilar although they are bisimilar. According to their reachability graphs in Fig. 7.3, we easily obtain a bisimulation:

$$\{(M_j, M_j'), (M_j', M_j) \mid j \in \mathbb{N}_{10}\}.$$

However, it is not a secure bisimulation since M_5 is insecure but M_5' is secure.

As mentioned above, the determinism/nondeterminism of the *internal* actions of a system can affect the interaction between the system and its user, and thus the notion of (weak) bisimulation can distinguish the two pattens of interactions. The above example shows that some *external* events (especially those unobservable and nondeterministic events that do not belong to the system) can also affect the interaction when they happen in some states of a system (e.g. tampering with a data). These states are defined as *insecure*. Therefore, it is reasonable to partition the states of a system into two parts.

Theorem 7.2 ([5, 6]) $|\widetilde{\approx}|$ *is an equivalence relation.*

Proof This conclusion is guaranteed by the following three cases:

- (Reflexivity) $(N^{is}, M_0) |\widetilde{\approx}| (N^{is}, M_0)$ for each (N^{is}, M_0). It is obvious.
- (Symmetry) $(N_1^{is}, M_{0_1}) |\widetilde{\approx}| (N_2^{is}, M_{0_2})$ means $(N_2^{is}, M_{0_2}) |\widetilde{\approx}| (N_1^{is}, M_{0_1})$; and vice versa. This is guaranteed by the symmetry of each secure bisimulation.
- (Transitivity) If $(N_1^{is}, M_{0_1}) |\widetilde{\approx}| (N_2^{is}, M_{0_2})$ and $(N_2^{is}, M_{0_2}) |\widetilde{\approx}| (N_3^{is}, M_{0_3})$, then we have $(N_1^{is}, M_{0_1}) |\widetilde{\approx}| (N_3^{is}, M_{0_3})$. Since $(N_1^{is}, M_{0_1}) |\widetilde{\approx}| (N_2^{is}, M_{0_2})$, there is a secure bisimulation $|\widetilde{\approx}|_1$ between (N_1^{is}, M_{0_1}) and (N_2^{is}, M_{0_2}); since $(N_2^{is}, M_{0_2}) |\widetilde{\approx}| (N_3^{is}, M_{0_3})$, there is a secure bisimulation $|\widetilde{\approx}|_2$ between (N_2^{is}, M_{0_2}) and (N_3^{is}, M_{0_3}). Now we can construct another binary relation as follows:

$$\{(M, M'), (M', M) \mid \exists M'' \in R(N_2^{is}, M_{0_2}) : M |\widetilde{\approx}|_1 M'' \wedge M'' |\widetilde{\approx}|_2 M'\}.$$

Obviously, this binary relation is a secure bisimulation and (M_{0_1}, M_{0_3}) is in it. Therefore, $(N_1^{is}, M_{0_1}) |\widetilde{\approx}| (N_3^{is}, M_{0_3})$.   □

Is it impossible that (weak) bisimulation does not distinguish the two systems in Fig. 7.2?

Actually, if we add the external events into them, we can also utilise the weak bisimulation to distinguish them. As shown in Fig. 7.4, the transitions t_{11} and t_{12}

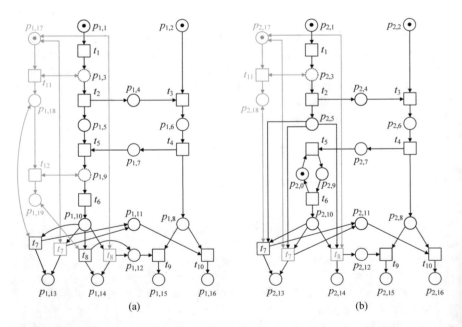

Fig. 7.4 The Petri nets modelling the electronic funds transfer system in which the actions of the Trojan program are considered. For simplification, a self-loop is represented by an arc with arrowheads at both ends

are thought of as two unobservable actions. They represent that the Trojan program tampers with the data items, and they are actually unobservable for both the user and the bank system. Additionally, the two Petri nets should be viewed as a labelled system, i.e., two transitions have the same label, representing the same action. For example, the transition t_8 in red and the transition t_8 in green in Fig. 7.4 both represent the user enters the verification code and agrees with this transfer, but firing the red t_8 means that the money is transferred into Account 3. Similarly, the transition t_7 in red and the transition t_7 in green in Fig. 7.4 both represent the user cancels this transfer. Although firing the red t_7 possibly means that the user finds an abnormality from the verification message and thus cancels this transfer and firing the green t_7 possibly means that the user does not want to continue this transfer and thus cancels it, they are both implemented in this system via clicking the button *cancel*.

Figure 7.5 shows the reachability graphs of the two Petri nets in Fig. 7.4, where ε represents unobservable actions. Through the two reachability graphs, we can see that there exists no weak bisimulation between them, and thus the two Petri nets in Fig. 7.4 are not equivalent. In Fig. 7.5a, the marking in red means that the money is transferred into Account 3, but there is no such a marking in (b), which means that the security of the system modelled by the Petri net in Fig. 7.2b is higher than that in Fig. 7.2a.

After we have a sense of observation and un-observation from the above example, we can define weak secure bisimulation where the transition set is divided into two parts: *observable* and *unobservable*. It is also worthy noting that for such a Petri net, some transitions have the same labels as shown in Fig. 7.4, i.e., it should be defined as a *labelled Petri net* [5, 6], but we do not take this definition here for simplicity.

Definition 7.5 (*Weak Secure Bisimulation of Two Petri Nets with Insecure Places*) Let $(N_1^{is}, M_{0_1}) = (P_{S_1} \cup P_{I_1}, T_O \cup T_U, F_1, M_{0_1})$ and $(N_2^{is}, M_{0_2}) = (P_{S_2} \cup P_{I_2}, T_O \cup T_U, F_2, M_{0_2})$ be two Petri nets with insecure places where T_O is the set of observable transitions, T_U is the set of unobservable transitions, and $(P_{S_1} \cup P_{I_1}) \cap (P_{S_2} \cup P_{I_2}) = \emptyset$. Let $\mathbb{IM}_1$ and $\mathbb{IM}_2$ be their insecure marking sets. A binary relation

$$\mathrel{\overset{\approx}{\imath\jmath}} \subseteq (R(N_1^{is}, M_{0_1}) \times R(N_2^{is}, M_{0_2})) \cup (R(N_2^{is}, M_{0_2}) \times R(N_1^{is}, M_{0_1}))$$

is a *weak secure bisimulation* if the following four conditions hold:

1. $\mathrel{\overset{\approx}{\imath\jmath}}$ is symmetric,
2. if $M_1 \mathrel{\overset{\approx}{\imath\jmath}} M_2$, $M_1[t\rangle M_1'$, and $t \in T_U$, then there exists M_2' and $\sigma \in T_U^*$ such that $M_2[\sigma\rangle M_2'$ and $M_1' \mathrel{\overset{\approx}{\imath\jmath}} M_2'$,
3. if $M_1 \mathrel{\overset{\approx}{\imath\jmath}} M_2$, $M_1[t\rangle M_1'$, and $t \in T_O$, then there exists M_2' and $\sigma \in T_U^*$ such that $M_2[\sigma t\rangle M_2'$ and $M_1' \mathrel{\overset{\approx}{\imath\jmath}} M_2'$,
4. if $M_1 \mathrel{\overset{\approx}{\imath\jmath}} M_2$, then M_1 and M_2 are either in $\mathbb{IM}_1 \cup \mathbb{IM}_2$ or not in $\mathbb{IM}_1 \cup \mathbb{IM}_2$.

If there is a weak secure bisimulation $\mathrel{\overset{\approx}{\imath\jmath}}$ such that $M_{0_1} \mathrel{\overset{\approx}{\imath\jmath}} M_{0_2}$, (N_1^{is}, M_{0_1}) and (N_2^{is}, M_{0_2}) are called *weakly securely bisimilar*, denoted as $(N_1^{is}, M_{0_1}) \mathrel{\overset{\approx}{\imath\jmath}} (N_2^{is}, M_{0_2})$.

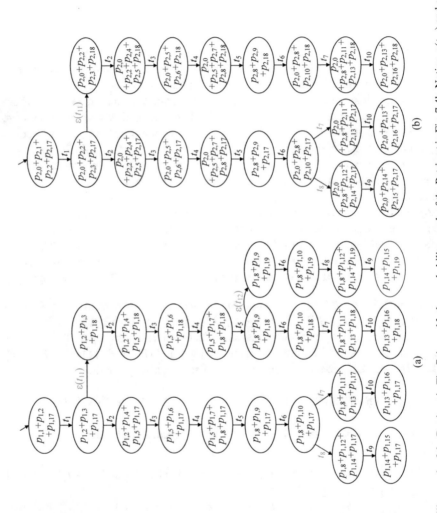

Fig. 7.5 **a** The reachability graph of the Petri net in Fig. 7.4a, and **b** the reachability graph of the Petri net in Fig. 7.4b. Notice, $\varepsilon(t_{11})$ and $\varepsilon(t_{12})$ are unobservable actions

Similarly, we have the following conclusion:

Theorem 7.3 ([5, 6]) $\approx$ is an equivalence relation.

? Question

Try to use knowledge-oriented Petri nets to model the systems mentioned above, use CTLK to represent the concerned properties, and analyse them.

7.3 Summery and Further Reading

Bisimulation improves our understanding to systems' behaviours and brings about a profound thinking to interaction and nondeterminacy [7–11]. If we may say that bisimulation and weak bisimulation consider the interaction between a system and its users in comparison with the classical language equivalence, then secure bisimulation considers the influence of external nondeterministic elements to those normal interactions. Certainly, this is a relatively primary consideration, while a further thinking is worthy, especially focusing on those cyber-physical systems as well as those open networked systems.

In addition to the language equivalence and the bisimulation equivalence, another kind of evaluation method of behaviour equivalence of systems is *behaviour consistence degree* computed according to a so-called *behaviour profiles* of systems. Some related knowledge can be found in [12–15].

References

1. Milner, R.: *Communication and Concurrency*. Prentice-Hall International, Englewood Cliffs (1989)
2. Milner, R.: *Communicating and Mobile Systems: the π-Calculus*. Cambridge University Press, UK (1999)
3. Hopcroft, J.E., Ullman, J.D.: *Introduction to Automata Theory, Languages and Computation*. Addison-Wesley, New York (1979)
4. Cassandras C.G., Lafortune, S.: *Introduction to Discrete Event Systems* (Third Edition). Springer Nature Switzerland AG (2021)
5. Liu, G.J., Jiang, C.J.: Secure bisimulation for interactive systems. In: *the 15th International Conference on Algorithms and Architectures for Parallel Processing*, pp. 625–639 (2015)
6. Liu, G.J, Jiang, C.J.: Behavioral equivalence of security-oriented interactive systems. *IEICE Transactions on Information and Systems* E99-D: 2061–2068 (2016)
7. Noroozi, A.A., Karimpour, J., Isazadeh, A.: Bisimulation for secure information flow analysis of multi-threaded programs. *Mathematical and Computational Applications* 24, article no. 64, 17 pages (2019)
8. Farhat, H.: Control of nondeterministic systems for bisimulation equivalence under partial information. *IEEE Transactions on Automatic Control* 65: 5437–5443 (2020)

9. Qin, X., Deng, Y., Du, W.: Verifying quantum Communication protocols with ground bisimulation. In: *the 26th International Conference on Tools and Algorithms for the Construction and Analysis of Systems*, pp. 21–38 (2020)

10. Belardinelli, F., Condurache, R., Dima, C., Jamroga, W., Knapik, M.: Bisimulations for verifying strategic abilities with an application to the ThreeBallot voting protocol. *Information and Computation*276, article no. 104552, 24 pages (2021)

11. Lanotte, R., Tini, S.: A weak semantic approach to bisimulation metrics in models with non-determinism and continuous state spaces. *Theoretical Computer Science* 869: 29–61 (2021)

12. Weidlich, M., Mendling, J., Weske, M.: Efficient consistency measurement based on behavioral profiles of process models. *IEEE Transactions on Software Engineering* 37: 410–429 (2010).

13. Wang, M., Liu, G.J., Zhao, P., Yan, C., Jiang, C.: Behavior consistency computation for workflow nets with unknown correspondence. *IEEE/CAA Journal of Automatica Sinica* 5: 281–291 (2017)

14. Wang, M., Ding, Z., Liu, G.J., Jiang, C., Zhou, M.C.: Measurement and computation of profile similarity of workflow nets based on behavioral relation matrix. *IEEE Transactions on Systems, Man, and Cybernetics: Systems* 50: 3628–3645 (2020)

15. Zhao, F., Xiang, D., Liu, G.J., Jiang, C.: A New Method for Measuring the Behavioral Consistency Degree of WF-Net Systems. *IEEE Transactions on Computational Social Systems* 9: 480–493 (2022)

Chapter 8
Time Petri Nets and Time-Soundness

8.1 Time Petri Nets

This section recalls the basic definitions, firing rules, and state class graphs of time Petri nets. More details can be found in [1–3].

8.1.1 Formal Definition and Firing Rules

We use $\mathbb{Q}$ to stand for the set of all nonnegative rational numbers and $\mathbb{R}$ the set of all nonnegative real numbers, although they usually stand for the sets of all rational and real numbers in lots of mathematical literatures, respectively.

Definition 8.1 (*Time Petri Net*) A *time Petri net* is denoted as $(N^t, M_0) \triangleq (P, T, F, I, M_0)$ where

1. (P, T, F, M_0) is a Petri net, and
2. $I: T \to \mathbb{Q} \times (\mathbb{Q} \cup \{\infty\})$ is a *static interval function* associated with all transitions such that for each $t \in T$:
$$\underline{I}(t) \leq \overline{I}(t).$$

8.1.1.1 Static Firing Interval and Newly Enabled Transitions

The notation
$$I(t) \triangleq \left[\underline{I}(t) \, .. \, \overline{I}(t)\right]$$

represents the *static firing interval* of the transition t. The lower bound $\underline{I}(t)$ and the upper bound $\overline{I}(t)$ stand for the earliest firing time and the latest firing time of

© The Author(s), under exclusive license to Springer Nature Singapore Pte Ltd. 2022
G. Liu, *Petri Nets*,
https://doi.org/10.1007/978-981-19-6309-4_8

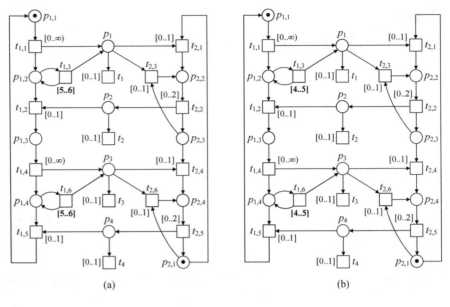

Fig. 8.1 Two time Petri nets modelling the alternating bit protocol [6–8]. The static firing interval of a transition "time-out and retransmitting" is [5..6] in (**a**) but [4..5] in (**b**)

t, respectively.[1] Obviously, a time Petri net is constructed through labelling a static firing interval for every transition of a Petri net.

Example 8.1 Figure 8.1 shows two time Petri nets whose difference lies in the static firing intervals of the transitions $t_{1,3}$ and $t_{1,6}$. The two time Petri nets model the *alternating bit protocol* [6–8]. In this protocol, a transmitter transmits messages to a receiver, and the latter transmits an acknowledgement to the former after receiving a message. But the transmission channel is unreliable, i.e., messages or acknowledgments may be lost in the channel. To recover from losses, a mechanism called *time-out and retransmitting* is used: after sending a message, the transmitter records the time, and if the related acknowledgment does not return within a given time, this message is retransmitted. Additionally, the receiver does not know whether the next message it receives is a new one or the copy of the last one it received. To solve this problem, a message is numbered with a modulo-2 sequence number. The messages with the two different sequence numbers are transmitted by turns. The transmission of a message with the sequence number 0 is modelled by transitions $t_{1,1}$, $t_{1,2}$, $t_{1,3}$, $t_{2,1}$, $t_{2,2}$, $t_{2,3}$, t_1, and t_2 in Fig. 8.1. The transmission of a message with the sequence number 1 is modelled by transitions $t_{1,4}$, $t_{1,5}$, $t_{1,6}$, $t_{2,4}$, $t_{2,5}$, $t_{2,6}$, t_3, and t_4. Notice that

---

[1] A *closed interval*, an *open interval*, and a *half-open interval* are usually represented by $[x..y]$, $(x..y)$, and $[x..y)$ or $(x..y]$, respectively, suggested by Hoare and Ramshaw [5]. Here, we use a closed interval to represent a static firing interval, except the case of $y = \infty$. In addition, two endpoints of a static firing interval are nonnegative rational numbers, but this interval is actually a set of real numbers.

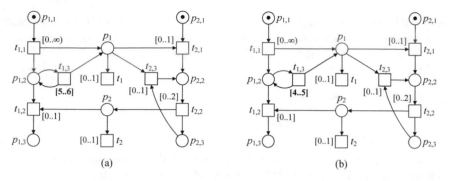

Fig. 8.2 The time Petri nets corresponding to half of the ones in Fig. 8.1

transitions $t_{1,3}$ and $t_{1,6}$ stand for the action *time-out and retransmitting* in the two transmission stages, and their time intervals are both [5..6] in Fig. 8.1a but [4..5] in (b). Firing t_1, t_2, t_3, or t_4 means the loss of a message or an acknowledgment in the transmission channel.

For simplicity, we use the time Petri nets in Fig. 8.2 instead of the ones in Fig. 8.1 as our running examples in what follows. They represent the transmission of a message with the sequence number 0.

We denote

$$En(M) \triangleq \{t \in T \mid M[t\rangle\}$$

as the set of all transitions enabled at M and

$$NEn(M, t) \triangleq \{t' \in En(M) \mid (t' = t) \vee (\exists p \in {}^\bullet t' \cap t^\bullet : M(p) = 1\}$$

as the set of all newly enabled transitions at M where M is reached by firing t.

> We may imagine that every transition in a time Petri net has a clock, and the clock of a transition starts to tick from 0 once this transition becomes enabled. After firing a transition, its clock either restores to 0 and stops working (if it becomes disabled) or starts to tick from 0 again (if it is enabled again).

Therefore, the definition of $NEn(M, t)$ is critical: the clock of a newly enabled transition starts to tick from 0. In this definition it is worthy noting the following cases:

– If t is still enabled at M, then it is a newly enabled transition at M.
– If another transition, say t', was not enabled before firing t but becomes enabled after firing t, then t' is a newly enabled transition at M.

– If another transition, say t', was enabled before firing t, and if there is an input
  place of t' such that it is both an input and an output of t and had exactly one token
  before firing t, then t' is also a newly enabled transition at M.
– There is no other case of "newly enabled".

Example 8.2 We assume that

$$M = p_{1,2} + p_{2,2} + p_1$$

is a marking of the Petri net in Fig. 8.2a and the static firing intervals of all transitions
are not considered temporally.[2] Firing $t_{1,3}$ at M yields the marking

$$M' = p_{1,2} + p_{2,2} + 2p_1.$$

Then we can obtain

$$En(M') = \{t_{1,3}, t_{2,2}, t_1\} \text{ and } NEn(M', t_{1,3}) = \{t_{1,3}\}.$$

8.1.1.2   State and Firing Rule

Definition 8.2 (*State of Time Petri Net*) A couple (M, h) is a *state* of a time Petri
net if M is a marking and the *clock function*

$$h : En(M) \to \mathbb{R}$$

satisfies that for each $t \in En(M)$:

$$h(t) \leq \overline{I(t)}.$$

We denote $s_0 = (M_0, h_0)$ as the *initial state* of a time Petri net where M_0 is the
initial marking and $h_0(t) = 0$ for all $t \in En(M_0)$.

Definition 8.3 (*State Transfer*) Let $s = (M, h)$ and $s' = (M', h')$ be two states of a
time Petri net. Then, there exist the following two rules of state transfer:

1. $s \xrightarrow{\tau} s'$ if and only if s' is reachable from s by the elapsing of time $\tau \in \mathbb{R}$, i.e.,

$$\begin{cases} M' = M \\ \forall t \in En(M') : h'(t) = h(t) + \tau \leq \overline{I(t)} \end{cases}.$$

2. $s \xrightarrow{t} s'$ if and only if s' is immediately reachable from s by the firing of the
   transition $t \in T$, i.e.,

---

[2] In fact, if the static firing interval of the transition $t_{1,3}$ is set up very small, e.g. $[0 \ldots 0.5]$, the
marking $M = p_{1,2} + p_{2,2} + p_1$ can be reached.

$$\begin{cases} t \in En(M) \\ M[t\rangle M' \\ \underline{I(t)} \le h(t) \le \overline{I(t)} \\ \forall t' \in NEn(M', t): h'(t') = 0 \\ \forall t' \in En(M') \setminus NEn(M', t): h'(t') = h(t') \end{cases}$$

When the clock value of an enabled transition is equal to or bigger than the lower bound of its static time interval, this transition may be fired, but it must be fired no later than the upper bound unless it becomes disabled due to the firing of other transitions.

Example 8.3 For the Petri net in Fig. 8.2a, the following state transfers are legal:

$$(p_{1,1} + p_{2,1}, h(t_{1,1}) = 0)$$
$$\downarrow 10$$
$$(p_{1,1} + p_{2,1}, h(t_{1,1}) = 10)$$
$$\downarrow t_{1,1}$$
$$(p_{1,2} + p_{2,1} + p_1, h(t_{1,3}) = 0 \wedge h(t_{2,1}) = 0 \wedge h(t_1) = 0)$$
$$\downarrow 0.8$$
$$(p_{1,2} + p_{2,1} + p_1, h(t_{1,3}) = 0.8 \wedge h(t_{2,1}) = 0.8 \wedge h(t_1) = 0.8)$$
$$\downarrow t_{2,1}$$
$$(p_{1,2} + p_{2,2}, h(t_{1,3}) = 0.8 \wedge h(t_{2,2}) = 0).$$

But, the following state transfer is illegal:

$$(p_{1,2} + p_{2,1} + p_1, h(t_{1,3}) = 0 \wedge h(t_{2,1}) = 0 \wedge h(t_1) = 0)$$
$$\downarrow 1.5$$
$$(p_{1,2} + p_{2,1} + p_1, h(t_{1,3}) = 1.5 \wedge h(t_{2,1}) = 1.5 \wedge h(t_1) = 1.5),$$

because when 1 unit of time elapsed at the state

$$(p_{1,2} + p_{2,1} + p_1, h(t_{1,3}) = 0 \wedge h(t_{2,1}) = 0 \wedge h(t_1) = 0),$$

one of $t_{2,1}$ and t_1 must be fired, so that it is not allowed that 1.5 units of time have elapsed but no transition has been fired.

8.1.2 Generation of State Class Graph

Generally, a time Petri net has infinitely many reachable states even when the number of its markings is finite. This is because there are infinitely many time points in a firable time interval and every time point corresponds to a different state. Because a firable time interval is a continual real set, the number of these states is not only infinite but also as dense as the real number set, i.e., it is uncountable. Therefore, it seems impossible to use a state-by-state way to represent and analyse the behaviours of a time Petri net. Fortunately, the technique of *state class graph* is feasible [1].

8.1.2.1 Clock Domain and State Class

Roughly speaking, a state class graph is an abstraction of the state space of a time Petri net where all states, which are reachable from the initial state by firing the same transition sequence, are grouped into a *state class*. A state class is characterised by two elements: one is the common marking of these states, and another is the union of *clock domains* for all enabled transitions.

A state class is represented by (M, C) where M is the common marking and C is the union of all clocks' domains for the enabled transitions.

Notice that C stands for all possible values of clocks associated with the enabled transitions when the corresponding marking is reached, i.e., each solution of C is a set of values of clocks indexed by the enabled transitions when the marking is reached.

The *initial state class* is denoted by $\pi_0 = (M_0, C_0)$ where M_0 is the initial marking and the initial clock domain is

$$C_0 = \bigwedge_{t \in En(M_0)} \left(h(t) = 0 \right),$$

i.e., it only includes the initial state.

For convenience, we use $R(N^t, M_0)$, $R(N^t, s_0)$, and $R(N^t, \pi_0)$ to represent the sets of reachable markings, reachable states, and reachable state classes of a time Petri net (N^t, M_0), respectively. $R(N^t, s_0)$ and $R(N^t, \pi_0)$ are usually called the *concrete state space* and *state class space* of (N^t, M_0), respectively [4].

8.1.2.2 Schedulability and Dynamic Firing Interval

A transition t is *schedulable* at a state s if there is time $\tau \in \mathbb{R}$ and two reachable states s' and s'' such that

$$s \xrightarrow{\tau} s' \wedge s' \xrightarrow{t} s''.$$

For convenience, we use

$$s \xrightarrow{(\tau,t)} s'' \text{ or } s \xrightarrow{\tau} s' \xrightarrow{t} s''$$

to stand for the above expression. Similarly, a transition sequence $\sigma = t_1 t_2 \ldots t_k$ is *schedulable* at a state s if there are states $s_1, s_2, \ldots, s_k$ and times $\tau_1, \tau_2, \ldots, \tau_k$ such that

$$s \xrightarrow{(\tau_1,t_1)} s_1 \xrightarrow{(\tau_2,t_2)} s_2 \ldots s_{k-1} \xrightarrow{(\tau_k,t_k)} s_k,$$

which is briefly denoted as

$$s \xrightarrow{(\mathcal{T},\sigma)} s_k$$

where $\mathcal{T}$ stands for the time sequence $\tau_1 \tau_2 \ldots \tau_k$.
  The following conclusion is obvious:

Lemma 8.1 ([3]) *Let $s \xrightarrow{(\mathcal{T}_1,\sigma_1)} s_1$ and $s \xrightarrow{(\mathcal{T}_2,\sigma_2)} s_2$ where $s_1 = (M_1, h_1)$ and $s_2 = (M_2, h_2)$. If $\psi(\sigma_1) = \psi(\sigma_2)$, then $M_1 = M_2$.*

  The procedure of generating the state class graph of a time Petri net is similar to that of generating a reachability graph of a Petri net. Given a newly yielded state class and a transition, we first decide if this transition is schedulable at this state class. If it is schedulable, then we compute the next state class. We repeat this process until no new state class occurs.
  Algorithm 8.1 is to decide if a transition t is schedulable at a given state class $\pi = (M, C)$.
  In fact, t is *schedulable* at π if and only if t is enabled at M and there exists $\tau \in \mathbb{R}$ and a state $s = (M, h) \in \pi$ such that t can be immediately fired at s' where

$$s \xrightarrow{\tau} s'.$$

The transition t can be immediately fired at s' if and only if

$$\left(\underline{I(t)} \le h(t) + \tau \right) \wedge \left(\bigwedge_{t' \in En(M)} \left(h(t') + \tau \le \overline{I(t')} \right) \right),$$

which obviously means that the value of the clock of t is in its static firing interval and the values of the clocks of other enabled transitions are not beyond their latest firing time.

Algorithm 8.1: IsSchedulable(π, t): Deciding If a Transition Is Schedulable at a State Class.

Input: A state class $\pi = (M, C)$ and a transition t;
Output: **true** (if t is schedulable at π) or **false** (otherwise);

1 **if** $t \notin En(M)$ **then**
2 | **return false**;

3 **else**
4 | **if** *there is* $\tau \in \mathbb{R}$ *and* $h \in C$ *such that*

$$\left(\left(\underline{I(t)} \leq h(t) + \tau \right) \right) \wedge \left(\bigwedge_{t' \in En(M)} \left(h(t') + \tau \leq \overline{I(t')} \right) \right)$$

    then
5 | | **return true**;
6 | **else**
7 | | **return false**;

Therefore, we should compute all possible τ under the constraint C and obtain a so-called *dynamic firing interval* $\theta_{(t,\pi)}$ of t at π. For simplicity, we write $\theta_{(t,\pi)}$ as θ_t when no ambiguity is introduced.

This dynamic firing interval can be computed by eliminating $h(t')$ of all transitions $t' \in En(M)$ from the following formula

$$\left(\theta_t \geq 0 \right) \wedge \left(\theta_t + h(t) \geq \underline{I(t)} \right) \wedge \left(\bigwedge_{t' \in En(M)} \left(\theta_t + h(t') \leq \overline{I(t')} \right) \right)$$

and guaranteeing this formula **true** under the given constraint C.

For example, $(p_{1,1} + p_{2,1}, h(t_{1,1}) = 0)$ is the initial state class of the time Petri net in Fig. 8.2b. Only $t_{1,1}$ is enabled, and thus its dynamic firing interval at this class can be computed according to the formula

$$(\theta_{t_{1,1}} \geq 0) \wedge (\theta_{t_{1,1}} + h(t_{1,1}) \leq \infty)$$

under the constraint $h(t_{1,1}) = 0$. After eliminating $h(t_{1,1})$, we have $\theta_{t_{1,1}} \geq 0$.

We take into account a more complex example of the state class

$$(p_{1,2} + p_{2,3} + p_2, h(t_{1,2}) = 0 \wedge 0 \leq h(t_{1,3}) \leq 3 \wedge h(t_2) = 0)$$

of the time Petri net in Fig. 8.2b. This state class means that the receiver sends the transmitter an acknowledgement. Therefore, the transitions $t_{1,2}$ and t_2 are enabled and their clocks are both 0's. Because the receiver spent $0 (= 0 + 0)$ to $3 (= 1 + 2)$ units of time to receive ($t_{2,1}$) a message and to send ($t_{2,2}$) an acknowledgement, the clock domain of the transition $t_{1,3}$ is $0 \le h(t_{1,3}) \le 3$. Later, after we introduce how to compute the successor of a state class, one can understand how this state class is generated. Here we first introduce how to compute the dynamic firing intervals of the three enabled transitions.

Example 8.4 To compute $\theta_{t_{1,2}}$, we first have the following formula:

$$(\theta_{t_{1,2}} \ge 0) \wedge (0 \le \theta_{t_{1,2}} + h(t_{1,2}) \le 1) \wedge (\theta_{t_{1,2}} + h(t_{1,3}) \le 5) \wedge (\theta_{t_{1,2}} + h(t_2) \le 1).$$

According to the given constraint $h(t_{1,2}) = h(t_2) = 0$, we know that $\theta_{t_{1,2}}$ should satisfies

$$(0 \le \theta_{t_{1,2}} \le 1) \wedge (\theta_{t_{1,2}} + h(t_{1,3}) \le 5).$$

In order to delete $h(t_{1,3})$ out of the above formula, we need another constraint $0 \le h(t_{1,3}) \le 3$. As a result, we have

$$0 \le \theta_{t_{1,2}} \le 1.$$

Similarly, we can compute the dynamic firing interval of t_2 at this state class is

$$0 \le \theta_{t_2} \le 1.$$

Example 8.5 To compute $\theta_{t_{1,3}}$, we first have the following formula:

$$(\theta_{t_{1,3}} \ge 0) \wedge (4 \le \theta_{t_{1,3}} + h(t_{1,3}) \le 5) \wedge (\theta_{t_{1,3}} + h(t_{1,2}) \le 1) \wedge (\theta_{t_{1,3}} + h(t_2) \le 1).$$

According to $h(t_{1,2}) = h(t_2) = 0$, we know that $\theta_{t_{1,3}}$ should satisfies

$$(\theta_{t_{1,3}} \ge 0) \wedge (4 \le \theta_{t_{1,3}} + h(t_{1,3}) \le 5) \wedge (\theta_{t_{1,3}} \le 1),$$

i.e.,

$$(0 \le \theta_{t_{1,3}} \le 1) \wedge (4 \le \theta_{t_{1,3}} + h(t_{1,3}) \le 5).$$

Due to $0 \le h(t_{1,3}) \le 3$, we finally compute the dynamic firing interval of $t_{1,3}$ at this state class is

$$1 \le \theta_{t_{1,3}} \le 1.$$

Algorithm 8.2: SUCCESSOR(π, t): Computing the Succeeding State Class of a Given One by Firing a Transition.

Input: A state class $\pi = (M, C)$ and a schedulable transition t;
Output: The succeeding state class of π via firing t;
1 Compute M' via $M[t\rangle M'$;
2 Compute the dynamic firing interval θ_t of t at π;
3 Replace $h(t')$ in C with $h(t') - \theta_t$ for each $t' \in En(M)$ and the result is denoted as C_1;
4 Construct a new constraint as follows:

$$C_2 \leftarrow \bigwedge_{t' \in En(M)} \left(h(t') \leq \overline{I(t')} \right) \wedge \left(\underline{I(t)} \leq h(t) \right) \wedge \theta_t \wedge C_1;$$

5 Eliminate θ_t from C_2 and denote the result as C_3;
6 Eliminate from C_3 all clock variables associated with transitions in $En(M) \setminus (En(M') \setminus NEn(M', t))$ and the result is denoted as C_4;
7 Construct a new constraint as follows:

$$C' \leftarrow \bigwedge_{t' \in NEn(M', t)} \left(h(t') = 0 \right) \wedge C_4;$$

8 **return** (M', C');

8.1.2.3   Computing Successors of State Classes and Generating State Class Graphs

Because there is a succeeding state reached via the elapsing of τ units of time for each $\tau \in \theta_t$ and at this state the transition t may be fired immediately, the clocks of other transitions have also run τ units of time when this transition is fired. Therefore, to compute π's successor yielded by firing t, we first make all clocks in π run θ_t units of time, i.e., we replace $h(t')$ in C with $h(t') - \theta_t$ for each $t' \in En(M)$, as shown in Line 3 in Algorithm 8.2. The value of the clock of every enabled transition cannot be beyond the latest firing time of this transition, and the value of the clock of t cannot be less than its earliest firing time because t is scheduled to fire. Consequently, we should carry out the operation in Line 4 in Algorithm 8.2. Furthermore, we remove the clocks of those transitions disabled at the new marking and add the clocks of those newly enabled ones, as shown in Lines 6 and 7 in Algorithm 8.2.

Therefore, according to the dynamic firing interval θ_t of a schedulable transition t at a state class π, π's successor yielded by firing t can be computed, as shown in Algorithm 8.2.

Example 8.6   According to the dynamic firing interval $0 \leq \theta_{t_{1,2}} \leq 1$ in Example 8.4, we use Algorithm 8.2 to compute the successor of the following state class

$$(p_{1,2} + p_{2,3} + p_2, h(t_{1,2}) = 0 \wedge 0 \leq h(t_{1,3}) \leq 3 \wedge h(t_2) = 0)$$

when $t_{1,2}$ in Fig. 8.2b is fired at this state class. In view of Line 3 in Algorithm 8.2, we have

$$(h(t_{1,2}) - \theta_{t_{1,2}} = 0) \wedge (0 \leq h(t_{1,3}) - \theta_{t_{1,2}} \leq 3) \wedge (h(t_2) - \theta_{t_{1,2}} = 0).$$

In view of Line 4, we have

$$(0 \leq h(t_{1,2}) \leq 1) \wedge (h(t_2) \leq 1) \wedge (h(t_{1,3}) \leq 5) \wedge$$

$$(h(t_{1,2}) = h(t_2) = \theta_{t_{1,2}}) \wedge (\theta_{t_{1,2}} \leq h(t_{1,3}) \leq 3 + \theta_{t_{1,2}}) \wedge (0 \leq \theta_{t_{1,2}} \leq 1).$$

According to Line 5, we eliminate $\theta_{t_{1,2}}$ from the above formula and then have

$$(0 \leq h(t_{1,2}) = h(t_2) \leq 1) \wedge (0 \leq h(t_{1,3}) \leq 4).$$

According to Lines 6 and 7, we remove the clocks of the disabled transitions and add the clocks of newly enabled transitions after firing $t_{1,2}$. After firing $t_{1,2}$, only $t_{1,4}$ is enabled. In other words, $h(t_{1,2})$, $h(t_2)$, and $h(t_{1,3})$ are all deleted from the new clock domains. All the above computations were wasted! Obviously, this successor is $(p_{1,3} + p_{2,3}, h(t_{1,4}) = 0)$. In fact, the above computations were not wasted from the aspect of understanding the system in Fig. 8.2b. From them we can see the changes of clock values of $t_{1,2}$, $t_{1,3}$, and t_2; and we can find that there exist the following clock values

$$h(t_{1,3}) = 4 \wedge h(t_{1,2}) = h(t_2) = 1$$

at which the transition $t_{1,3}$ can also be fired immediately. In other words, when the transmitter receives an acknowledgement (which means that the receiver has received the related message) and should have sent the next message, this transmitter may also resend the prior message, which means that there are some flaws in this system.

Example 8.7 We continue to consider the computation of the successor of the state class in Example 8.6 when firing the transition $t_{1,3}$. In Example 8.5, we have obtained its dynamic firing interval $\theta_{t_{1,3}} = 1$. According to the clock domains of the current state class and Line 3 of Algorithm 8.2, we have

$$(h(t_{1,2}) - \theta_{t_{1,3}} = 0) \wedge (0 \leq h(t_{1,3}) - \theta_{t_{1,3}} \leq 3) \wedge (h(t_2) - \theta_{t_{1,3}} = 0).$$

According to Line 4 of Algorithm 8.2, we have

$$(h(t_{1,2}) \leq 1) \wedge (h(t_2) \leq 1) \wedge (4 \leq h(t_{1,3}) \leq 5) \wedge$$

$$(h(t_{1,2}) = h(t_2) = \theta_{t_{1,3}}) \wedge (\theta_{t_{1,3}} \leq h(t_{1,3}) \leq 3 + \theta_{t_{1,3}}) \wedge (\theta_{t_{1,3}} = 1).$$

After eliminating $\theta_{t_{1,3}}$ from the above formula, we obtain the following clock domains:

$$h(t_{1,2}) = h(t_2) = 1 \wedge h(t_{1,3}) = 4.$$

Since firing $t_{1,3}$ yields the marking

$$p_{1,2} + p_{1,3} + p_1 + p_2$$

at which the transitions $t_{1,2}$, $t_{1,3}$, $t_{2,3}$, and t_1 are newly enabled and the transition t_2 keeps enabled. Therefore, we obtain the clock domains of this new state class as follows:

$$(h(t_{1,2}) = h(t_{1,3}) = h(t_{2,3}) = h(t_1) = 0) \wedge (h(t_2) = 1).$$

Algorithm 8.3 describes the procedure of generating the state class graph of a time Petri net. In this algorithm, $SCG(N^t, M_0)$ represents the state class graph of the time Petri net (N^t, M_0), $R(N^t, \pi_0)$ represents the set of all reachable state classes, and E represents the set of all edges.

Figure 8.3a and b show the state class graphs of the time Petri nets in Fig. 8.2a and b, respectively. First, we can see from Fig. 8.3 that different static firing intervals result in different state class graphs, which means that the two systems in Fig. 8.2 have different behaviours. For example, at the state class

$$(p_2 + p_{1,2} + p_{2,3}, h(t_2) = h(t_{1,2}) = 0 \wedge 0 \le h(t_{1,3}) \le 3),$$

the transition $t_{1,3}$ is schedulable in the time Petri net in Fig. 8.2b but is not schedulable in Fig. 8.2a. Because the static firing interval of $t_{1,3}$ is [4..5] in Fig. 8.2b, its clock may reach its earliest firing time 4, and thus at the above state class there exists an outgoing edge associated with $t_{1,3}$ in Fig. 8.3b; but in Fig. 8.2a, this case cannot take place since the earliest firing time of $t_{1,3}$ is 5.

On the other hand, from Fig. 8.3b we can see an abnormal phenomenon:

> The transmitter retransmits a message even when the receiver has received this message and the transmitter has received a related acknowledgment.

This abnormal phenomenon is exactly illustrated by the state class

$$(p_1 + p_{1,3} + p_{2,3}, h(t_1) = 0 \wedge h(t_{2,3}) = 0)$$

in Fig. 8.3b. It is caused just due to the nondeterminacy of behaviours of this system, while this nondeterminacy is caused by the inappropriate configurations of the static firing interval of $t_{1,3}$. In other words, [4..5] is inappropriate, while [5..6] is appropriate. In the next section, we will formalise this kind of (non)determinacy and propose a method to check it.

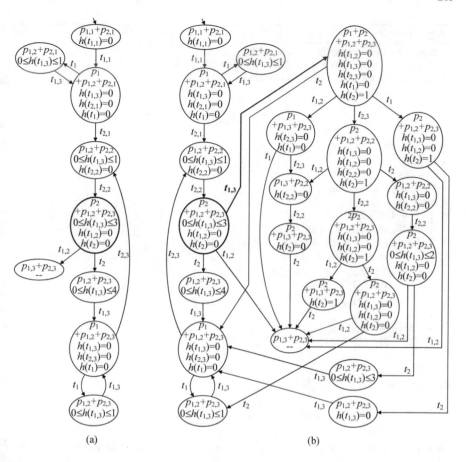

Fig. 8.3 The state class graphs corresponding to the two time Petri nets in Fig. 8.2, respectively

8.1.3 Completeness of State Class Graph

The following conclusions guarantee the correctness of these algorithms and illustrate that each reachable state can be represented in a state class graph and there is no pseudo state in the state class graph.

Lemma 8.2 ([1, 3]) *Let π be a state class in the state class graph of a time Petri net, t be schedulable at π, and θ_t be the dynamic firing interval of t at π. Then, for each value $\tau \in \theta_t$, there is a state $s \in \pi$ such that t can be immediately fired at such a state that is reached from s by the elapsing of time τ.*

Proof According to Line 4 of Algorithm 8.1 as well as the definition of dynamic firing interval, we know that for each $\tau \in \theta_t$, there is a state

$$s = (M, h) \in \pi = (M, C)$$

Algorithm 8.3: Producing the State Class Graph of a Time Petri Net.

Input: A time Petri net $(N^t, M_0) = (P, T, F, \mathcal{I}, M_0)$;
Output: The state class graph $SCG(N^t, M_0) = (R(N^t, \pi_0), E)$;
1 Construct the initial clock domains as follows:

$$C_0 \leftarrow \bigwedge_{t \in En(M_0)} (h(t) = 0);$$

2 $Q_{new} \leftarrow \emptyset$;
3 $R(N^t, \pi_0) \leftarrow \emptyset$;
4 $E \leftarrow \emptyset$;
5 **PutTail**($Q_{new}, (M_0, C_0)$);
6 **while** $Q_{new} \neq \emptyset$ **do**
7    $\pi \leftarrow$ **GetHead**(Q_{new});
8    $R(N^t, \pi_0) \leftarrow R(N^t, \pi_0) \cup \{\pi\}$;
9   **for** *each* $t \in T$ *such that* **IsSchedulable**$(\pi, t) =$ **true do**
10     $\pi' \leftarrow$ **SUCCESSOR**(π, t);
11   **if** $\pi' \notin Q_{new} \cup R(N^t, M_0)$ **then**
12       $\lfloor$ **PutTail**(Q_{new}, π');
13     $E \leftarrow E \cup \{(\pi, \pi')_t\}$;

14 **return** $(R(N^t, \pi_0), E)$;

---

such that for each $t' \in En(M)$:

$$h(t') + \tau \leq \overline{\mathcal{I}(t')} \wedge \underline{\mathcal{I}(t)} \leq h(t).$$

Therefore, according to the first rule of state transfer, we know that this conclusion is true. $\quad\square$

The converse of the above conclusion is also true.

Lemma 8.3 ([1, 3]) *Let π be a state class in the state class graph of a time Petri net, t be schedulable at π, and θ_t be the dynamic firing interval of t at π. For each state $s \in \pi$ such that t can be fired immediately at the state that is reached from s by the elapsing of time τ, then we have*

$$\tau \in \theta_t.$$

Proof This is obvious according to the definition of dynamic firing interval. $\quad\square$

The above two conclusions mean that θ_t stands for all possible firing time of t at the state class π. Therefore, the state class, which is computed by **SUCCESSOR**(π, t) based on θ_t, fulfils the following two conclusions:

Lemma 8.4 ([1, 3]) *Let $\pi' =$ **SUCCESSOR**(π, t). Then, for each state $s' \in \pi'$, there is a state $s \in \pi$ and $\tau \in \theta_t$ such that*

$$s \xrightarrow{(\tau, t)} s'.$$

Lemma 8.5 ([1, 3]) *Let $\pi' = \text{SUCCESSOR}(\pi, t)$. Then, for each state $s \in \pi$, if there is $\tau \in \theta_t$ such that*

$$s \xrightarrow{(\tau, t)} s',$$

we have

$$s' \in \pi'.$$

Lemmata 8.2 and 8.4 mean that a state class graph does not contain any pseudo states, while Lemmata 8.3 and 8.5 mean that all reachable states are represented by a state class graph.

Theorem 8.1 *The state class graph of a time Petri net contains all reachable state, and no pseudo state is in it.*

Question

Try to analyse the complexities of these algorithms. Try to write a computer program of generating the state class graph of a time Petri net, especially a program of computing a dynamic firing interval as well as a successor. Is the state class graph finite when the number of markings of a time Petri net is finite?

8.2 Time-Soundness

This section first defines *time-soundness* and then shows that it can guarantee the determinacy of behaviours of a time Petri net in view of the notion *bisimulation*. Finally, an algorithm is designed to decide if a time Petri net is time-sound. For more details, one may read [3].

8.2.1 Time-Soundness Based on Bisimulation

8.2.1.1 Time-Soundness and Solo State Graph

Definition 8.4 (*Time-soundness*) A time Petri net $(N^t, M_0) = (P, T, F, I, M_0)$ is *time-sound* if for any transition $t \in T$ and any two reachable states s_1 and s_2 that satisfy

$$\exists \mathcal{T}_1 \in \mathbb{R}^*, \exists \mathcal{T}_2 \in \mathbb{R}^*, \exists \sigma \in T^*: s_0 \xrightarrow{(\mathcal{T}_1, \sigma)} s_1 \wedge s_0 \xrightarrow{(\mathcal{T}_2, \sigma)} s_2,$$

we have that t is schedulable at s_1 if and only if t is schedulable at s_2.

In this definition, the constraint on $s_1 = (M_1, h_1)$ and $s_2 = (M_2, h_2)$ means that s_1 and s_2 have the same marking (i.e., $M_1 = M_2$) by Lemma 8.1. That is, this marking can be reached via the same transition sequence (i.e., σ) at different times (h_1 may be different from h_2 due to the difference between $\mathcal{T}_1$ and $\mathcal{T}_2$). Time-soundness requires that for an arbitrary transition, it is either schedulable or not at all such states.

Time-soundness means that a system has deterministic behaviours always. To prove this, we first involve a so-called *solo state graph*.

Definition 8.5 (*Solo State Graph*) A solo state graph of a time Petri net $(N^t, M_0) = (P, T, F, \mathcal{I}, M_0)$ is a labelled digraph $SSG(N^t, s_0) \triangleq (V, E, R(N^t, s_0), L_V, L_E, v_0)$ where

1. V and E are the sets of nodes and directed edges, respectively,
2. $v_0 \in V$ is the initial node, and
3. $L_V: V \to R(N^t, s_0)$ and $L_E: E \to (\mathbb{R} \times T)$ are two label functions such that

   a. $L_V(v_0) = s_0$, and
   b. for each $v \in V$, if $t \in T$ is schedulable at $L_V(v)$, then there is exactly one time $\tau \in \mathbb{R}$ and exactly one node $v' \in V$ such that

   $$L_E(v, v') = (\tau, t) \wedge L_V(v) \xrightarrow{(\tau, t)} L_V(v').$$

A solo state graph requires that if a transition is schedulable at a state, then a feasible firing time for this transition is randomly chosen and the related reachable state is produced. Any other feasible firing time is not considered any more in this solo state graph. For example, Fig. 8.4a and b illustrate two solo state graphs of the time Petri net in Fig. 8.2a. From the randomness of choosing a firing time, we know that for any reachable state or any transition sequence from the initial state, we can always construct a solo state graph such that the state or the transition sequence is in it. Obviously, a time Petri net can have infinitely many solo state graphs and a solo state graph perhaps has infinite nodes.

Because any two solo state graphs of a time-sound time Petri net are bisimilar, we can assert that time-soundness ensures the determinacy of behaviours of a time-sound time Petri net. Before proving that any two solo state graphs of a time-sound

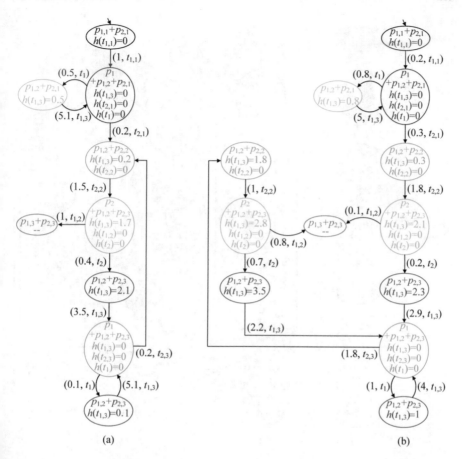

Fig. 8.4 Two solo state graphs of the time Petri net in Fig. 8.2a. They are bisimilar since there is a bisimulation as shown by the state pairs in the same colours

time Petri net are bisimilar, we first illustrate two solo state graphs of an un-time-sound time Petri net in order to understand the notion of time-soundness. Figure 8.5 shows two solo state graphs of the time Petri net in Fig. 8.2b. Obviously, they are not bisimilar because at the state

$$(p_2 + p_{1,2} + p_{2,3}, h(t_2) = h(t_{1,2}) = 0 \wedge h(t_{1,3}) = 2.4)$$

in Fig. 8.5a there are two schedulable transitions t_2 and $t_{1,2}$, at the state

$$(p_2 + p_{1,2} + p_{2,3}, h(t_2) = h(t_{1,2}) = 0 \wedge h(t_{1,3}) = 3)$$

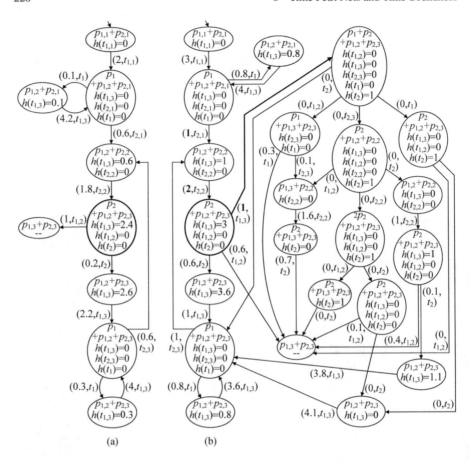

Fig. 8.5 Two solo state graphs of the time Petri net in Fig. 8.2b. They are not bisimilar

in Fig. 8.5b there are three schedulable transitions t_2, $t_{1,2}$, and $t_{1,3}$, whereas the two states may be reached by firing the same transition sequence $t_{1,1}t_{2,1}t_{2,2}$. Therefore, the time Petri net in Fig. 8.2b is not time-sound.

8.2.1.2   The Relation Between Time-Soundness and Bisimulation

When we define a bisimulation over two solo state graphs, the clock values of every state and the time label of every edge should be omitted. At this time, every solo state graph can be viewed as an automaton. For understanding it easily, we redefine bisimulation over two solo state graphs.

Definition 8.6 (*Bisimulation of Two Solo State Graphs*) Let $SSG(N^t, s_0)_j = (V_j, E_j, R(N^t, s_0), L_{V_j}, L_{E_j}, v_{0_j})$ be two solo state graphs of the time Petri net

$(N^t, M_0) = (P, T, F, \mathcal{I}, M_0)$, where $j = 1, 2$. $SSG(N^t, s_0)_1$ and $SSG(N^t, s_0)_2$ are *bisimilar*, denoted as $SSG(N^t, s_0)_1 \approx SSG(N^t, s_0)_2$, if there is a symmetric binary relation $\approx \subseteq (V_1 \times V_2) \cup (V_2 \times V_1)$ satisfying the following conditions:

1. $v_{0_1} \approx v_{0_2}$, and
2. for each $v_1 \approx v_2$, if $(v_1, v_1') \in E_1$ and $L_{E_1}(v_1, v_1') = (\tau_1, t)$, then there is $\tau_2 \in \mathbb{R}$ and $(v_2, v_2') \in E_2$ such that

$$L_{E_2}(v_2, v_2') = (\tau_2, t) \wedge v_1' \approx v_2'.$$

For example, the two solo state graphs in Fig. 8.4a and b are bisimilar where two nodes in the same colour are bisimilar.

Lemma 8.6 ([3]) *Any two solo state graphs of a time-sound time Petri net are bisimilar.*

Proof Let $SSG(N^t, s_0)_j = (V_j, E_j, R(N^t, s_0), L_{V_j}, L_{E_j}, v_{0_j})$ be two solo state graphs of the time Petri net $(N^t, M_0) = (P, T, F, \mathcal{I}, M_0)$, where $j = 1, 2$. Here, we first prove that for each directed path $v_0, v_1 v_2 \ldots v_k$ in $SSG(N^t, s_0)_1$, there always exists a directed path $v_{0_2} v_1' v_2' \ldots v_k'$ in $SSG(N^t, s_0)_2$ such that

- $L_{E_1}(v_{0_1}, v_1)_T = L_{E_2}(v_{0_2}, v_1')_T$
- $L_{E_1}(v_1, v_2)_T = L_{E_2}(v_1', v_2')_T$
- $\cdots$
- $L_{E_1}(v_{k-1}, v_k)_T = L_{E_2}(v_{k-1}', v_k')_T$

where $L_E(x, y)_T = t$ if and only if $L_E(x, y) = (\tau, t)$ for some τ. This can be proven by the following contradiction method.

We assume that in $SSG(N^t, s_0)_1$, there is a directed path that starts at v_{0_1} but does not satisfy the above condition. Then, we can find a directed path $v_{0_1} x_1 x_2 \ldots x_k$ in $SSG(N^t, s_0)_1$ satisfying the following two points:

1. for any directed path starting at v_{0_2} in $SSG(N^t, s_0)_2$ and having a length k, the transition sequence labeled on this path is not equal to the transition sequence labeled on $v_{0_1} x_1 x_2 \ldots x_k$, but
2. there exists a directed path $v_{0_2} y_1 y_2 \ldots y_k$ in $SSG(N^t, s_0)_2$ such that

$$L_{E_1}(v_{0_1}, x_1)_T = L_{E_2}(v_{0_2}, y_1)_T$$
$$L_{E_1}(x_1, x_2)_T = L_{E_2}(y_1, y_2)_T$$
$$\cdots$$
$$L_{E_1}(x_{k-2}, x_{k-1})_T = L_{E_2}(y_{k-2}, y_{k-1})_T$$
$$L_{E_1}(x_{k-1}, x_k)_T \neq L_{E_2}(y_{k-1}, y_k)_T.$$

From the second point and Lemma 8.1, we have that

$$L_{V_1}(x_{k-1}) = L_{V_2}(y_{k-1})$$

and they are reachable from the initial state. However, according to the definition of time-soundness we know that a transition is schedulable at $L_{V_1}(x_{k-1})$ if and only if this transition is schedulable at $L_{V_2}(y_{k-1})$. By the definition of solo state graph, we thus have that for the node x_k, there is a node y in $SSG(N^t, s_0)_2$ such that

$$L_{E_1}(x_{k-1}, x_k)_T = L_{E_2}(y_{k-1}, y)_T$$

because the transition $L_{E_1}(x_{k-1}, x_k)_T$ is also schedulable at $L_{V_2}(y_{k-1})$. This contradicts the first point.

From the above conclusion we have that for each node $v \in V_1$ and each directed path $v_0, v_1 v_2 \ldots v_k v$, there always exists a node $v' \in V_2$ and a directed path $v_{0_2} v_1' v_2' \ldots v_k' v'$ such that

- $L_{E_1}(v_{0_1}, v_1)_T = L_{E_2}(v_{0_2}, v_1')_T$
- $L_{E_1}(v_1, v_2)_T = L_{E_2}(v_1', v_2')_T$
- $\ldots$

- $L_{E_1}(v_k, v)_T = L_{E_2}(v_k', v')_T$,

and vice versa. Obviously, all such pairs (v, v') and (v', v) form a bisimulation. □

From the above proof we can see that two bisimilar states in two bisimilar solo state graphs have the same marking.

Lemma 8.7 ([3]) *A time Petri net is time-sound if any two solo state graphs of it are bisimilar.*

Proof (By contradiction) We assume that a time Petri net $(N^t, M_0) = (P, T, F, I, M_0)$ is not time-sound but any two solo state graphs of it are bisimilar. Then, there are two different states $s, s' \in R(N^t, s_0)$, two groups of times $\tau_{1_1}, \ldots, \tau_{1_k}, \tau_{2_1}, \ldots, \tau_{2_k} \in \mathbb{R}$, and a group of transitions $t_1, \ldots, t_k \in T$ such that

$$s_0 \xrightarrow{(\tau_{1_1}, t_1)} s_{1_1} \ldots s_{1_{k-1}} \xrightarrow{(\tau_{1_k}, t_k)} s_{1_k} = s \wedge s_0 \xrightarrow{(\tau_{2_1}, t_2)} s_{2_1} \ldots s_{2_{k-1}} \xrightarrow{(\tau_{2_k}, t_k)} s_{2_k} = s',$$

and there is a transition $t \in T$ such that t is schedulable at s but not schedulable at s'. We can construct two solo state graphs $SSG(N^t, s_0)_j = (V_j, E_j, R(N^t, s_0), L_{V_j}, L_{E_j}, v_{0_j})$ in both which there exists a directed path $v_{0_j} v_{j_1} v_{j_2} \ldots v_{j_k}$ such that

- $L_{V_j}(v_{0_j}) = s_0$
- $L_{V_j}(v_{j_1}) = s_{j_1}$
- $\ldots$

- $L_{V_j}(v_{j_k}) = s_{j_k}$
- $L_{E_j}(v_{0_j}, v_{j_1}) = (\tau_{j_1}, t_1)$
- $L_{E_j}(v_{j_1}, v_{j_2}) = (\tau_{j_2}, t_2)$

– ...

- $L_{E_j}(v_{j_{k-1}}, v_{j_k}) = (\tau_{j_k}, t_k)$

where $j \in \{1, 2\}$. Hence, we have that

$$v_{1_1} \approx v_{2_1}$$

due to

$$SSG(N^t, s_0)_1 \approx SSG(N^t, s_0)_2 \wedge v_{0_1} \approx v_{0_2} \wedge L_{E_1}(v_{0_1}, v_{1_1})_T = t_1 = L_{E_2}(v_{0_2}, v_{2_1})_T.$$

Accordingly, we know that

$$(v_{1_2} \approx v_{2_2}) \wedge \cdots \wedge (v_{1_k} \approx v_{2_k}).$$

Because t is schedulable at

$$s = s_{1_k} = L_{V_1}(v_{1_k}),$$

we know that $SSG(N^t, s_0)_1$ has a node v such that

$$L_{E_1}(v_{1_k}, v)_T = t.$$

Therefore, $SSG(N^t, s_0)_2$ should also have a node v' such that

$$L_{E_2}(v_{2_k}, v')_T = t$$

due to

$$v_{1_k} \approx v_{2_k}.$$

This means that t is also schedulable at

$$L_{V_2}(v_{2_k}) = s_{2_k} = s'.$$

Obviously, a contradiction is yielded.   □

From Lemmata 8.6 and 8.7 we have the following conclusion:

Theorem 8.2 ([3]) *A time Petri net is time-sound if and only if any two solo state graphs of it are bisimilar.*

Based on this conclusion we know that the time Petri net in Fig. 8.2b is not time-sound because it has two solo state graphs as shown in Fig. 8.5 which are not bisimilar. In theory, the number of solo state graphs of a time Petri net is possible infinite, and the number of nodes of a solo state graph may be infinite. Therefore, it is hard technically to decide the time-soundness of a time Petri net via solo state graphs. Next, we introduce a method of using state class graphs to perform this task.

8.2.2 Deciding Time-Soundness

The following conclusion guarantees that the state class graph of a time Petri net can be used to decide its soundness.

Theorem 8.3 *A time Petri net is time-sound if and only if for each state class of its state class graph and for each transition schedulable at this class, this transition is schedulable at each state of this class.*

Proof As stated in the previous lemmata, the state class graph of a time Petri net exactly contains all reachable states of it. According to the construction of a state class graph, we know that any two states that can be reached by firing the same transition sequence (via different time sequences) are grouped in the same state class. Therefore, the sufficiency obviously holds. As for the necessity, we assume that there is a state class and a transition such that this transition is schedulable at some states of this class but it is not schedulable at other states of this class. Obviously, this case violates the definition of time-soundness, which contradicts the pre-condition that the time Petri net is time-sound.   □

On the basis of this conclusion, we can utilise the state class graph to decide the time-soundness of a time Petri net, i.e., we only need to decide whether each transition that is schedulable at a state class is still schedulable at all states of this class. We can use the dynamic firing interval θ_t of a transition t schedulable at a state class $\pi = (M, C)$ to generate the maximum subclass of π such that t is schedulable at all states of this subclass. This subclass is denoted as $\pi_t = (M, C_t)$ and called *schedulable subclass of t at π.*

Obviously, we have the following conclusion:

Corollary 8.1 ([3]) *A time Petri net is time-sound if and only if $\pi_t = \pi$ holds for each state class π of its state class graph and each transition t schedulable at π.*

The computation of a schedulable subclass is similar to that of a successor, as shown in Algorithm 8.4. We travel a state class graph to perform **SubClass** for each node and each schedulable transition, and finally can determine the time-soundness. We omit the description of this deciding procedure, but in what follows we illustrate the procedure of computing schedulable subclasses through two examples.

Question

Why is $h(t') + \theta_t$ rather than $h(t') - \theta_t$ used in Line 1 of Algorithm 8.4?

Algorithm 8.4: SubClass(π, t): Compute the Schedulable Subclass of a Transition at a State Class.

Input: A state class $\pi = (M, C)$ and the dynamic firing interval θ_t of a transition t at π where t is schedulable at π;

Output: Schedulable subclass π_t;

1 Construct the following constraint:

$$C' \leftarrow \bigwedge_{t' \in En(M)} \left(h(t') + \theta_t \leq \overline{I(t')} \right) \wedge \left(\underline{I(t)} + \theta_t \leq h(t) \right) \wedge \theta_t \wedge C;$$

2 Eliminate θ_t from C' and denote the result by C_t;
3 $\pi_t \leftarrow (M, C_t)$;
4 **return** π_t;

Example 8.8 For the state class

$$(p_2 + p_{1,2} + p_{2,3}, h(t_{1,2}) = h(t_2) = 0 \wedge 0 \leq h(t_{1,3}) \leq 3)$$

in Fig. 8.3b, we have obtained the dynamic firing interval of the transition $t_{1,2}$ at this state class (see Example 8.4):

$$0 \leq \theta_{t_{1,2}} \leq 1.$$

To compute the schedulable subclass of $t_{1,2}$, according to Algorithm 8.4 we first construct the following constraint:

$$(0 \leq \theta_{t_{1,2}} \leq 1) \wedge (h(t_{1,2}) = h(t_2) = 0) \wedge (0 \leq h(t_{1,3}) \leq 3) \wedge$$

$$(0 \leq h(t_{1,2}) + \theta_{t_{1,2}} \leq 1) \wedge (h(t_2) + \theta_{t_{1,2}} \leq 1) \wedge (h(t_{1,3}) + \theta_{t_{1,2}} \leq 5).$$

After eliminating $\theta_{t_{1,2}}$ from the above constraint, we obtain the clock domain of the schedulable subclass of $t_{1,2}$ as follows:

$$h(t_{1,2}) = h(t_2) = 0 \wedge 0 \leq h(t_{1,3}) \leq 3,$$

which is identical to that of the original class.

Example 8.9 We continually consider the schedulable subclass of $t_{1,3}$ at the state class in the above example. We have known that its dynamic firing interval is $\theta_{t_{1,3}} = 1$ at this state class (see Example 8.5). Hence, to compute the schedulable subclass of $t_{1,3}$, we first construct the following constraint:

$$(\theta_{t_{1,3}} = 1) \wedge (h(t_{1,2}) = h(t_2) = 0) \wedge (0 \leq h(t_{1,3}) \leq 3) \wedge$$

$$(h(t_{1,2}) + \theta_{t_{1,3}} \leq 1) \wedge (h(t_2) + \theta_{t_{1,3}} \leq 1) \wedge (4 \leq h(t_{1,3}) + \theta_{t_{1,3}} \leq 5).$$

After eliminating $\theta_{t_{1,3}}$ from the above constraint, we obtain the clock domain of the schedulable subclass of $t_{1,3}$ as follows:

$$h(t_{1,2}) = h(t_2) = 0 \wedge h(t_{1,3}) = 1,$$

which is not identical to that of the original state class. From this schedulable subclass, we know that the time Petri net in Fig. 8.2b is not time-sound.

Similarly, we can check through the state class graph in Fig. 8.3a that every related schedulable subclass is equal to the original state class, which means that the time Petri net in Fig. 8.2a is time-sound.

8.3 Application

We have seen that the alternating bit protocol is correct when the static firing interval of the action "time-out and retransmitting" is [5..6], but has an potential error when this interval is [4..5]. Here, we consider another example: *level crossing* (LC) [9–12], that is often used as a benchmark in the fault diagnose field. We illustrate how to use the notion of time-soundness to check the safety of the control system of a *multi-track level crossing with sensors*.

An LC is an intersection where one or multiple railway lines cross a road at the same level. The control system of an LC is composed of multiple sensors, sound alarms, road lights, barriers and a controller. For example, the LC shown in Fig. 8.6a has two tracks and one road. Once a train enters into the *control zone*, a corresponding

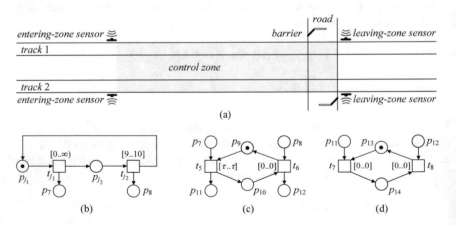

Fig. 8.6 a Schematic diagram of a two-track level crossing with four sensors, **b** the module of railway traffic, **c** the module of LC controller, and **d** the module of barriers

entering-zone sensor sends a signal to the *controller*. Then it immediately triggers the *sound alarms* and *road lights* informing the cars and people in the crossing to leave the crossing as soon as possible and will lower *barriers* in a fix period. Once the train leaves the control zone, a corresponding *leaving-zone sensor* sends a signal to the controller that will immediately trigger barriers to rise. Here we assume that a train spends 9–10 units of time to cross the control zone.

Here, the control system is modelled by a time Petri net that is combined by three subsystems: *railway traffic*, *LC controller* and *barriers*. Notice that we do not consider the sound alarms and road lights in order to simplify the model since the controls on them are similar to that on barriers. Next, we introduce the models of these subsystems that are similar to those given in [10].

Railway Traffic

The railway traffic with sensors can be modelled by the time Petri net in Fig. 8.6b.

- The marked place p_{j_1} (where $j \in \{1, 2\}$ is the track index) denotes that a train is approaching the control zone delimited by the jth entering-zone sensor and the jth leaving-zone sensor.
- Firing the transition t_{j_1} means that the jth entering-zone sensor detects that a train is entering the control zone. Then it sends a signal (p_7) to the controller.
- The marked place p_{j_2} denotes that the train is in the control zone.
- Firing the transition t_{j_2} means that the jth leaving-zone sensor detects that the train leaves the control zone. Then it also sends the controller a signal (p_8).

Notice that p_{j_1} is the output place of t_{j_2}, which means that a train is allowed to enter the control zone along the jth track only if the previous one has left the zone along this track.

The unbounded time interval $[0..\infty)$ associated with t_{j_1} implies that a train can entering the control zone at any time. The interval $[9..10]$ associated with t_{j_2} means that a train takes 9–10 units of time to pass through the zone.

LC Controller

The LC controller collects signals from entering-/leaving-zone sensors and sends controlling commands to sound alarms, road lights and barriers. Here we only consider the commands to barriers.

Figure 8.6c shows a controller given in [10]. A token in p_7 means that a train enters the control zone and a token in p_8 means that a train leaves the crossing. Obviously, firing t_5 implies that a command of closing barriers (p_{11}) is sent and firing t_6 is to send a command of opening barriers (p_{12}).

We first set up a time interval for t_5 and t_6, respectively. When the controller receives a signal from an entering-zone sensor, it cannot immediately send a command of closing barriers (but it may send commands of triggering alarm sounds and road lights). The reason is that some cars or people are possibly in the crossing at this time. Therefore, barriers should be closed after τ units of time after the controller receives the signal from the entering-zone sensor. Here, τ should be greater than 0

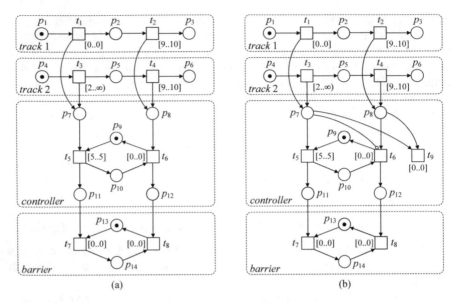

Fig. 8.7 **a** A time Petri net modelling a 2-track LC in which the controller is ineffective, and **b** a time Petri net modelling this system in which the controller is effective

and less than 9 ($\tau = 5$ in the following examples). It is reasonable to associate [0..0] with t_6 because the command of opening barriers can be immediately sent once the train leaves the crossing.

Obviously, such a controller is effective for one-track LC. However, for multi-track LC it can result in an un-time-sound system, i.e., an action sequence leads sometimes to a safe situation but sometimes to an unsafe one (e.g., while a train is running through the crossing, barriers are open). Later, such a case is illustrated.

Barriers

The time Petri net modelling the barrier system is shown in Fig. 8.6d. Barriers will be immediately lowered once a command of closing barriers (p_{11}) is reached and then they are in the "down" state (p_{14}). They will be immediately raised when a command of opening barriers (p_{12}) comes and then they are in the "up" state (p_{13}). Here the time intervals associated with t_7 and t_8 are both [0..0].

Analysis

Based on the above subsystems, we can construct a time Petri net model of the two-track LC, as shown in Fig. 8.7a. Notice that we simplify the model but this does not affects the illustration of the usefulness of time-soundness. Here, the time-soundness can be used to show that:

– the controller in Fig. 8.6c results in an error when two trains are both in the control zone in an overlapping period, and
– a correct controller is time-sound.

To model two parallel trains and make the model as simple as possible, we use two un-cycle time Petri nets (i.e., "track 1" and "track 2" in Fig. 8.7a) instead of the time Petri net with cycles in Fig. 8.6b. Additionally, the time interval [0..0] associated with t_1 means that the first train enters the control zone at the initial state of the model. The time interval [2..∞) associated with t_3 implies that the second train is allowed to enter the control zone after at least two units of time after the first train enters.

We consider the following two scheduling sequences in Fig. 8.7a as well as the yielded states:

– $(0, t_1)(5, t_5)(0, t_7)(0, t_3)(4, t_2)(0, t_6)(0, t_8)$ is schedulable at the initial state and it leads to the following state

$$(p_3 + p_5 + p_7 + p_9 + p_{13}, h(t_4) = 4 \wedge h(t_5) = 0).$$

– $(0, t_1)(5, t_5)(0, t_7)(4, t_3)(0, t_2)(0, t_6)(0, t_8)$ is also schedulable at the initial state, leading to the following state

$$(p_3 + p_5 + p_7 + p_9 + p_{13}, h(t_4) = h(t_5) = 0).$$

Obviously, transitions t_4 and t_5 are both schedulable after the first sequence is fired, but only transition t_5 is schedulable (but t_4 is not) after the second sequence is fired. Therefore, the time Petri net in Fig. 8.7a is not time-sound. The second case implies that barriers are closed after 5 units of time after the second train enters the control zone. The first case means the following unsafe situation:

Barriers are re-opened after 4 units of time after the second train enters the zone, and they remain opening while this train passes through the crossing.

This is because after barriers are re-opened, the second train will pass through the crossing in 5–6 units of time but the command of re-closing barriers will be sent in 5 units of time.

In a word, this controller is incorrect since it cannot safely control a multiple-track LC. Figure 8.8 clearly shows the state class graph of the Petri net in Fig. 8.7a. From Fig. 8.8 we can clearly see that there exist non-deterministic behaviours at those two state classes in bold ellipses. For example of the state class

$$(p_3 + p_5 + p_7 + p_9 + p_{13}, 4 \leq h(t_4) \leq 8 \wedge h(t_5) = 0),$$

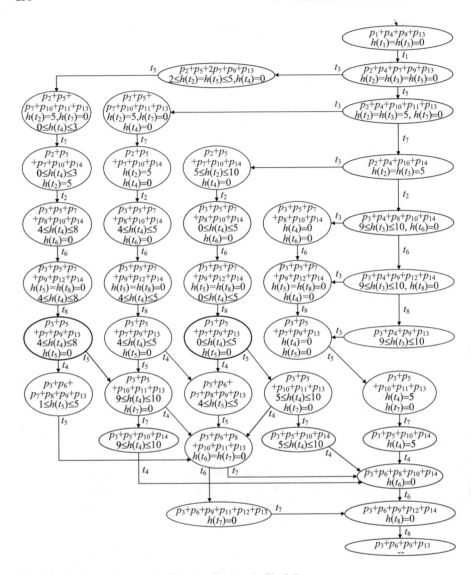

Fig. 8.8 The state class graph of the time Petri net in Fig. 8.7a

at its subclass

$$(p_3 + p_5 + p_7 + p_9 + p_{13}, 4 \le h(t_4) \le 5 \wedge h(t_5) = 0),$$

the transition t_5 can be scheduled to fire, but at the following subclass

$$(p_3 + p_5 + p_7 + p_9 + p_{13}, 5 < h(t_4) \le 8 \wedge h(t_5) = 0),$$

it cannot be scheduled to fire.

A Correct Controller

In what follows, we design an effective controller based on the above one. The idea is that the controller must check if there is still a train in the control zone when it receives a signal from a leaving-zone sensor. In other words, it can send a command of opening barriers only if there is no train in the zone. If there is another train in the zone, it will deactivate this signal. Figure 8.7b illustrates this controller in which

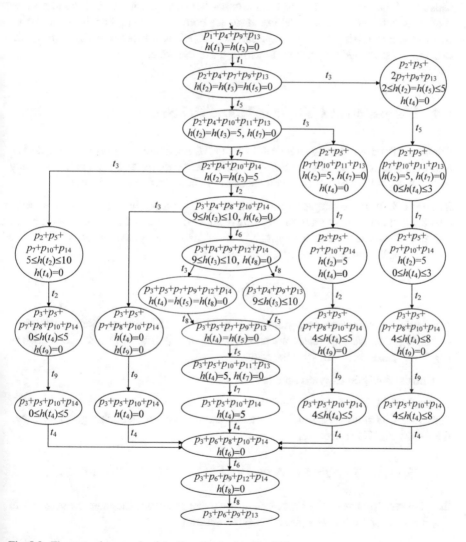

Fig. 8.9 The state class graph of the time Petri net in Fig. 8.7b

transitions t_6 and t_9 work for these checks.[3] Figure 8.9 shows the state class graph of the time Petri net in Fig. 8.7b. We can check through this state class graph that the time Petri net in Fig. 8.7b is time-sound.

The two transition sequences

$$t_1 t_5 t_7 t_2 t_6 t_8 t_3 t_5 t_7 t_4 t_6 t_8 \text{ and } t_1 t_5 t_7 t_2 t_6 t_3 t_8 t_5 t_7 t_4 t_6 t_8$$

illustrate the case that only one train crosses the control zone at a time, and other sequences indicate the cases that two train are both in the zone. For those cases that two trains are passing through the zone in parallel, it is clear that the barriers are ordered to be opened only when they both leave the zone.

8.4   Discussion of CTL over Time Petri Nets

For a time Petri net, we may construct some CTL formulae to represent some design requirements and use its state class graph to verify them; and especially, we may omit the clock domains of this state class graph in a verification procedure.

Example 8.10   For example of the alternating bit protocol in Fig. 8.2, if we want to check whether the phenomenon "the transmitter both received an acknowledgement and re-transmitted a related message" can occur in this protocol, we can use the following CTL formula to represent it:

$$\mathbf{EF}\Big(p_1 \wedge p_{1,3} \Big).$$

And based on their state class graphs in Fig. 8.3, especially without considering their clock domains, we can verify the above formula. It is true in Fig. 8.3a but false in (b). These are coincident with the facts.

Unfortunately, such good luck is rare.

Example 8.11   For the time Petri net in Fig. 8.7a, if we do not consider the clock domains of its state class graph in Fig. 8.8 and use such a reachable graph to check the following CTL formula:

$$\mathbf{AG}\Big(\big(p_3 \wedge p_5 \wedge p_7 \wedge p_9 \wedge p_{13} \big) \Rightarrow \mathbf{EF}\big(p_3 \wedge p_5 \wedge p_{10} \wedge p_{11} \wedge p_{13} \big)\Big),$$

then the checking result is **true**, because we can see from the state class graph in Fig. 8.8 that for each state class π with the marking

---

[3] For simplification we use an inhibitor arc in this model. It is easy to define the enabling/firing rules of transitions of time Petri nets with inhibitor arcs [13, 14], and their time-soundness and state class graphs are the same with those in this chapter. In fact, we can also use some complementary places to eliminate this inhibitor arc. Readers interested in it may try to design it.

$$p_3 + p_5 + p_7 + p_9 + p_{13},$$

there is a state class π' with the marking

$$p_3 + p_5 + p_{10} + p_{11} + p_{13}$$

such that there exists a directed edge from π to π'. We want to use this formula to represent such an assertion that when the marking

$$p_3 + p_5 + p_7 + p_9 + p_{13}$$

is reached, there exists a run leading to the marking

$$p_3 + p_5 + p_{10} + p_{11} + p_{13}.$$

In fact, this formula is **not always true** in the original system. As shown in Sect. 8.3, the state class

$$(p_3 + p_5 + p_7 + p_9 + p_{13}, 4 \leq h(t_4) \leq 8 \wedge h(t_5) = 0)$$

has a subclass

$$(p_3 + p_5 + p_7 + p_9 + p_{13}, 5 < h(t_4) \leq 8 \wedge h(t_5) = 0),$$

and the transition t_5 cannot be scheduled to fire at each state of this subclass, which means that the marking

$$p_3 + p_5 + p_{10} + p_{11} + p_{13}$$

cannot be reached from any state of this subclass. In other words, when the system enters a state of this subclass, any succeeding run cannot lead to the marking

$$p_3 + p_5 + p_{10} + p_{11} + p_{13}.$$

This exactly illustrates the characteristics of an un-time-sound time Petri net.

Since a time-sound time Petri net has deterministic behaviours, we can verify CTL formulas through its state class graph and the verification procedure can omit the clock domains of all state classes. In other words, the state class graph of a time-sound time Petri net can be viewed as a deterministic automaton, while the clock domain of each state class can be omitted. Theorem 8.3 and Corollary 8.1 can guarantee all these.

However, if a time Petri net is not time-sound and a CTL formula is checked through its state class graph without considering its clock domains, then the checking result is not necessarily correct, as shown in the above example.

In a word, it is infeasible that a state class graph of an un-time-sound time Petri net is directly thought of as a reachable graph without clock domains so as to check CTL formulae.

Even for a time-sound one, CTL is sometimes incapable of representing what we want to verify.

For example of the level crossing in Fig. 8.7b, when we want to verify whether the barrier is unsafely opened while the second train is passing through the crossing, it seems that the following CTL formula can represent this design requirement:

$$\mathbf{EF}\Big(\mathbf{E}\big((p_5 \wedge p_{14})\mathbf{U}(p_5 \wedge p_{13})\big)\Big).$$

This formula means that there is a running path in which p_5 and p_{14} have held (i.e., the second train is in the zone and the barrier is in the "down" state) until p_5 and p_{13} hold (i.e., the second train is still in the zone but the barrier is opened). Although the time Petri net in Fig. 8.7b is time-sound, we use its state class graph in Fig. 8.9 to check this formula and find that it is indeed **true** due to the existence of such a path

$$p_3 + p_5 + p_7 + p_9 + p_{12} + p_{14} \xrightarrow{t_8} p_3 + p_5 + p_7 + p_9 + p_{13}.$$

Is this controlled system incorrect? But we have said in Sect. 8.3 that it is correct.

In fact, this system requires that in five units of time after a train enters the control zone, the barrier may be in the open state. The above running path exactly indicates that when the first train just leaves the control zone and the controller has sent the signal of opening the barrier but the barrier has not been opened, the second train enters the control zone; and then the barrier is opened; but after five units of time, it will be closed again; hence, this controlled system has no problem.

What we want to express through this example is that the time factor should be considered in the temporal formulae so as to represent some more complex design requirements.

When clock domains have to be taken into account and some design requirements pay attention to time constraints, the TCTL and its verification emerge and will be introduced in the next chapter.

8.5   Summery and Further Reading

This chapter introduces the basic knowledge of time Petri nets as well as a property, namely *time-soundness*, which aims at representing the determinacy of behaviours of some time-critical systems. The idea of naming "time-soundness" comes from the soundness of workflow nets. Time-soundness means that for each state class π and each transition t schedulable at π, t is schedulable at each state in π (see Theorem 8.3 and Corollary 8.1).

From the examples in this chapter, e.g. Fig. 8.8, we can see that some different state classes correspond to the same marking; therefore, there have been some studies on how to group some of them so as to save storage space [15–17]; however, these grouping methods usually generate pseudo states. In addition to the technique of state class graph, there are other techniques such as *geometric region graph* [18] and *zone based graph* [19, 20] to analyse time Petri nets. For more knowledge of time Petri nets, one can read [21, 22]. TINA is a famous tool of time Petri nets [23].

With the development of technology, real-time systems become more and more popular and important. However, most of them are composed of physical processes and computing processes, and time plays a critical role in their correctness. All of them bring more challenges for a correct design. Formal modelling of these systems is an important way to analyse them [24–26]. However, it is not enough for their correctness analysis only to consider the time factor. There are many other factors affecting the correctness of such a complex system, such as scheduling policy, interrupting mechanism, and priority. Chapter 10 will introduce the Petri-net-based modelling and analysis from the aspects of these factors.

References

1. Hadjidj, R., Boucheneb, H.: Improving state class constructions for CTL* model checking of time Petri nets. *International Journal of Software Tools Technology Transfer* 10: 167–184 (2008)
2. Bachmann, J.P., Popova-Zeugmann, L.: Time-independent liveness in time Petri nets. *Fundamenta Informaticae* 101: 1–17 (2010)
3. Liu, G.J., Jiang, C.J., Zhou, M.C.: Time-soundness of time Petri nets Modelling time-critical systems. *ACM Transactions on Cyber-Physical Systems* 2, article number 11, 27 pages (2018)
4. Boucheneb, H., Hadjidj, R.: CTL* model checking for time Petri nets. *Theoretical Computer Science* 353: 208–227 (2006)
5. Graham, R.L., Knuth, D.E., Patashnik, O.: *Concrete Mathematics: A Foundation for Computer Science*. Addison-Wesley Publishing Company (2002)
6. Merlin, P., Faber, D.J.: Recoverability of communication protocols - implications of a theoretical study. *IEEE Transactions on Communications* 24: 1036–1043 (1976)
7. Berthomieu, B., M. Diaz, M.: Modeling and verification of time dependent systems using time Petri nets. *IEEE Transactions on Software Engineering* 17: 259–273 (1991)
8. Jbeli, N., Sbaï, Z., Ben Ayed, R. (2018). TCTL$_h^\Delta$ Model Checking of Time Petri Nets. *Lecture Notes in Computer Science* 11120: 242–262 (2018)
9. Leveson, G., Stolzy, J.L.: Safety analysis using Petri nets. *IEEE Transactions on Software Engineering* 13: 386–397 (1987)
10. Ghazel, M., Liu, B.: A customizable railway benchmark to deal with fault diagnos is issues in DES. In: *the 13th International Workshop on Discrete Event Systems*, pp. 177–182 (2016)
11. Wu, D., Zheng, W.: Formal model-based quantitative safety analysis using timed Coloured Petri Nets. *Reliability Engineering and System Safety* 176: 62–79 (2018)
12. Fayyaz, M.A.B., Alexoulis-Chrysovergis, A.C., Southgate, M.J., Johnson, C.: A review of the technological developments for interlocking at level crossing. *Proceedings of the Institution of Mechanical Engineers, Part F: Journal of Rail and Rapid Transit* 235: 529–539 (2021)
13. Roux, O.H., Lime, D.: Time Petri nets with inhibitor hyperarcs: Formal semantics and state space computation. In: *the 25th International Conference on Application and Theory of Petri Nets*, pp. 371–390 (2004)

14. Pencolé, Y., Subias, A.: Diagnosability of event patterns in safe labeled time Petri nets: a model-checking approach. *IEEE Transactions on Automation Science and Engineering* 19: 1151–1162 (2022)
15. Berthomieu, B., Vernadat, F.: State class constructions for branching analysis of time Petri nets. In: *the 9th International Conference on Tools and Algorithms for the Construction and Analysis of Systems*, pp. 442–457 (2003)
16. Lubat, É., Dal Zilio, S., Botlan, D. L., Pencolé, Y., Subias, A.: A state class construction for computing the intersection of time Petri nets languages. In: *the 17th International Conference on Formal Modeling and Analysis of Timed Systems* pp. 79-95 (2019)
17. Jbeli, N., Sbai, Z.: On improving model checking of time Petri nets and its application to the formal verification. *International Journal of Service Science, Management, Engineering, and Technology* 12: 68–84 (2021)
18. Yoneda, T., Ryuba, H.: CTL model checking of time Petri nets using geometric regions. *IEICE Transactions on Information and Systems* 81: 297–306 (1998)
19. Gardey, G., Roux, O. H., Roux, O.F.: Using zone graph method for computing the state space of a time Petri net. In: *the 1st International Conference on Formal Modeling and Analysis of Timed Systems*, pp. 246–259 (2003)
20. Parrot, R., Briday, M., Roux, O.H.: Timed Petri nets with reset for pipelined synchronous circuit design. In: *the 42th International Conference on Applications and Theory of Petri Nets and Concurrency*, pp. 55–75 (2021)
21. Wang, J.C.: *Timed Petri nets: Theory and Application*. Springer Science and Business Media (2012)
22. Popova-Zeugmann, L.: *Time and Petri Nets*. Springer, Berlin (2013)
23. Berthomieu, B., Vernadat, F.: Time Petri nets analysis with TINA. In: QEST 6: 123– 124 (2006) (https://projects.laas.fr/tina/index.php)
24. Nicolescu, G., Mosterman, P.J.: *Model-based Design for Embedded Systems*. CRC Press (2018)
25. Platzer, A.: *Logical Foundations of Cyber-Physical Systems*. Springer, Cham (2018)
26. Taha, W.M., Taha, A.E.M., Thunberg, J.: *Cyber-Physical Systems: A Model-Based Approach*. Springer Nature (2021)

Chapter 9
Verifying Timed Computation Tree Logic Based on Time Petri Nets

9.1 Syntax and Semantics of TCTL

This section introduces the syntax and semantics of *timed computation tree logic* (TCTL). For more similar work, one may read [1].

9.1.1 Syntax of TCTL

We first assume that a time Petri net considered here has finitely many state classes.

Given a time Petri net $(N^t, M_0) = (P, T, F, \mathcal{I}, M_0)$, all places are still viewed as atomic propositions.

Given a state $s_1 \in R(N^t, s_0)$, $\Omega(s_1)$ denotes all *computation paths* starting at s_1. A *computation path with finitely many positions*

$$s_1 \xrightarrow{(\tau_1, t_1)} s_2 \xrightarrow{(\tau_2, t_2)} s_3 \cdots x_{k-1} \xrightarrow{(\tau_{k-1}, t_{k-1})} s_k$$

means that no transition is schedulable at s_k. A *computation path with infinitely many positions*

$$s_1 \xrightarrow{(\tau_1, t_1)} s_2 \xrightarrow{(\tau_2, t_2)} s_3 \cdots x_{k-1} \xrightarrow{(\tau_{k-1}, t_{k-1})} s_k \cdots$$

means that there is at least one schedulable transition at every state in this path.

In addition to the starting state of a given computation path, every state yielded by the firing of a transition rather than by the elapsing of time corresponds to a *position* of this path. The set of all positions of a computation path ω is denoted as $positions(\omega)$.

Formally speaking, although the notation

$$s_1 \xrightarrow{(\tau_1, t_1)} s_2$$

means that there is a state s' such that

$$s_1 \xrightarrow{\tau_1} s' \xrightarrow{t_1} s_2,$$

the state s' does not correspond to a position of this computation path.

Given a computation path ω and a position $j \in positions(\omega)$, $\omega(j)$ represents the state lying in the j-th position of this path. For example of the above computation paths, the states lying in the first, second, and third positions are s_1, s_2, and s_3, respectively.

For convenience, given a computation path ω and two positions $j - 1$ and j, we denote

$$\omega(j - 1) \xrightarrow{(\tau_{j-1}^{\omega}, t_{j-1}^{\omega})} \omega(j).$$

According to the first firing rule of time Petri nets, a new state is yielded via elapsing τ units of time from $\omega(j - 1)$ for each $\tau \leq \tau_{j-1}^{\omega}$, and this state also occurs in this computation path. For convenience, we denote this new state as

$$\omega(j - 1)[\tau\rangle$$

instead of the following form

$$\omega(j - 1) + \tau$$

often used in other literatures.

Notice that $\omega(j - 1)$ and $\omega(j - 1)[\tau\rangle$ have the same marking but different clock values for each $\tau \leq \tau_{j-1}^{\omega}$ except the case of $\tau = 0$, i.e., $\omega(j - 1) = \omega(j - 1)[0\rangle$. In this chapter, any time variable like τ is greater than or equal to 0 by default.

For a computation path with finitely many positions (say k positions), it has no τ_k^{ω} since the state lying in the k-th position is a deadlock, but we assume $\tau_k^{\omega} = 0$. This assumption will bring convenience for the description of TCTL semantics.

Definition 9.1 (*Syntax of TCTL*) Let AP be the set of atomic propositions and $\mathbf{I} \in \mathbb{Q} \times (\mathbb{Q} \cup \{\infty\})$ is a time interval. Then, TCTL formulae can be defined according to the following grammar:

$$\phi ::= \mathbf{true} \mid \mathbf{deadlock} \mid p \mid \neg\phi \mid \phi_1 \wedge \phi_2 \mid \mathbf{EG_I}\, \phi \mid \mathbf{E}(\phi_1\, \mathbf{U_I}\, \phi_2)$$

where $p \in AP$, $\mathbf{true} \notin AP$, and $\mathbf{deadlock} \notin AP$.

In TCTL [2], there is no *next* operator because time is dense and thus the next state cannot be defined. The time interval $\mathbf{I}$ in a TCTL formula may be closed, or open, or half-open.

9.1.2 Semantics of TCTL

Definition 9.2 (*Semantics of TCTL*) Let $s = (M, h)$ and $p \in AP$. The relation $\models$ is defined inductively as follows:

- $s \models$ **true**.
- $s \models$ **deadlock** if and only if no transition is schedulable at s.
- $s \models p$ if and only if $M(p) > 0$.
- $s \not\models \phi$ if and only if $s \models \neg\phi$.
- $s \models \phi_1 \wedge \phi_2$ if and only if $s \models \phi_1$ and $s \models \phi_2$.
- $s \models \mathbf{E}(\phi_1 \; \mathbf{U_I} \; \phi_2)$ if and only if there exists $\omega \in \Omega(s)$, $j \in positions(\omega)$, and $\tau \leq \tau_j^\omega$ such that

$$
\begin{cases}
\sum_{k=1}^{j-1} \tau_k^\omega + \tau \in \mathbf{I} \\
\omega(j)[\tau\rangle \models \phi_2 \\
\forall k \in \mathbb{N}_{j-1}^+, \forall \tau' \leq \tau_k^\omega : \omega(k)[\tau'\rangle \models \phi_1 \\
\forall \tau' < \tau : \omega(j)[\tau'\rangle \models \phi_1
\end{cases}
$$

- $s \models \mathbf{A}(\phi_1 \; \mathbf{U_I} \; \phi_2)$ if and only if for each $\omega \in \Omega(s)$ there exists $j \in positions(\omega)$ and $\tau \leq \tau_j^\omega$ such that

$$
\begin{cases}
\sum_{k=1}^{j-1} \tau_k^\omega + \tau \in \mathbf{I} \\
\omega(j)[\tau\rangle \models \phi_2 \\
\forall k \in \mathbb{N}_{j-1}^+, \forall \tau' \leq \tau_k^\omega : \omega(k)[\tau'\rangle \models \phi_1 \\
\forall \tau' < \tau : \omega(j)[\tau'\rangle \models \phi_1
\end{cases}
$$

$s \models \mathbf{E}(\phi_1 \; \mathbf{U_I} \; \phi_2)$ means that there exists a computation path ω starting at $s = \omega(1)$ such that ϕ_1 has held along this path until the occurrence of another state $\omega(j)[\tau\rangle$ satisfying ϕ_2, and the running time from the state s to the state $\omega(j)[\tau\rangle$ lies in the time interval $\mathbf{I}$. $s \models \mathbf{A}(\phi_1 \; \mathbf{U_I} \; \phi_2)$ means for all paths the above requirement is satisfied.

Other operators can be defined on the basis of $\mathbf{EU_I}$ or $\mathbf{AU_I}$ as follows:

- $\mathbf{EF_I} \; \phi \triangleq \mathbf{E}(\mathbf{true} \; \mathbf{U_I} \; \phi)$,
- $\mathbf{AF_I} \; \phi \triangleq \mathbf{A}(\mathbf{true} \; \mathbf{U_I} \; \phi)$,
- $\mathbf{EG_I} \; \phi \triangleq \neg(\mathbf{AF_I} \; \neg\phi)$,
- $\mathbf{AG_I} \; \phi \triangleq \neg(\mathbf{EF_I} \; \neg\phi)$.

Question

A CTL formula of **AU** can be transferred into a formula composed of **EG** and **EU**. Can a TCTL formula of **AU_I** be transferred into a similar form? If not, try to provide an explanation. According the definitions of $\mathbf{EF_I}$, $\mathbf{AF_I}$, $\mathbf{EG_I}$, and $\mathbf{AG_I}$ above, try to provide the semantics descriptions for them like Definition 9.2.

Example 9.1 In the example of level crossing in Chap. 8, it is permitted that the barrier is closed after 5 units of time after a train enters the control zone, and this train will spend 9–10 units of time to pass through the control zone. This implies that the barrier has been closed for 4–5 units of time until this train leaves this control zone. Therefore, we can use the following TCTL formula to represent this design requirement w.r.t. the second train:

$$\mathbf{AG}_{[0..\infty)}\left(\left(p_5 \wedge \neg p_{14}\right) \Rightarrow \mathbf{AF}_{[0..\infty)}\left(\mathbf{A}\left((p_5 \wedge p_{14})\mathbf{U}_{[4..5]}(\neg p_5 \wedge p_{14})\right)\right)\right).$$

This formula represents that if the second train enters the control zone but the barrier is still open, then for any future running path there is another state satisfying that starting at this state, this train has been in the control zone and the barrier has been closed for 4–5 units of time, until this train leaves.

It is not easy to design a backward labelling method like the ones in Chap. 5 to compute the satisfiable state set of an $\mathbf{EU_I}$ or $\mathbf{AU_I}$ formula, because it is not easy to compute the time intervals of these paths during a backward procedure or compute those satisfiable predecessors in a state class graph. In the next section, we introduce a forward method that utilises a depth-first strategy to find those potentially satisfiable computation paths in a state class graph. Based on these computation paths, we compute satisfiable states.

9.2 Verifying TCTL Based on State Class Graphs of Time Petri Nets

This section first describes the recursive procedure of computing the satisfiable state set of a TCTL formula, and then introduces the computations of satisfiable states of $\mathbf{EU_I}$ and $\mathbf{AU_I}$ formulae in detail.

9.2.1 Recursively Computing Satisfiable State Sets of TCTL Formulae

For a time Petri net with finitely many state classes, its state class graph represents all reachable states and no pseudo state is in the state class graph. Hence, we can utilise a state class graph to compute those states satisfying a TCTL formula.

We construct a 3-tuple $(\pi, C_{\tau_t}, \tau_t)$ to represent a set of satisfiable states where

- $\pi = (M, C)$ stands for a node (state class) of a state class graph,
- C_{τ_t} is called the *guided clock domains* w.r.t. a transition t such that (M, C_{τ_t}) as a subclass of π represents all satisfiable states of a formula at π such that t is schedulable at all states of (M, C_{τ_t}), and
- τ_t is the *guided dynamic firing interval* of a transition t at (M, C_{τ_t}) such that t is schedulable at all states of (M, C_{τ_t}).
For convenience, we name this 3-tuple structure as a *guided tuple*.

First, τ_t is a subset of θ_t, and C_{τ_t} is a subset of C. Additionally, according to π of a guided tuple, we easily carry out a depth-first traversal starting at this guided tuple in a state class graph. In each traversal procedure, if a node of a state class graph has k schedulable transitions, we may imagine that it is virtually (but actually not) partitioned into k nodes and each node is labelled by such a guided tuple in which the first element indicates the related node in this state class graph, the second element indicates all satisfiable states of some formula at this state class, and the third element indicates the related edge (labelled by t) in the state class graph as well as the dynamic firing interval of t at (M, C_{τ_t}). Notice, we may use Algorithm 8.4 in Chap. 8 to compute this subclass based on this guided dynamic firing interval.

Algorithm 9.1 describes this recursive computation procedure of the satisfiable state set of a TCTL formula according to different forms of its sub-formulae.
In what follows, we use the following TCTL formula

$$\mathbf{E}\Big((p_5 \wedge p_{13})\mathbf{U}_{[3.9..4.2]}(p_6 \wedge p_{13})\Big)$$

as an illustration example which describes a property of the time Petri net in Fig. 8.7 (a). This formula represents that there exists a computation path in which the barrier has been open for 3.9 to 4.2 units of time until the second train leaves the crossing.

Example 9.2 We first consider the satisfiable state set of the sub-formula $p_5 \wedge p_{13}$ that can be computed based on Lines 3, 4, 7, and 8 of Algorithm 9.1 and the state class graph in Fig. 8.8. These satisfiable states are listed in Table 9.1, represented by guided tuples. For instance of the state class

$$(p_3 + p_5 + p_7 + p_9 + p_{13}, 4 \leq h(t_4) \leq 8 \wedge h(t_5) = 0),$$

Algorithm 9.1: SAT-T(ϕ): Recursively Computing the Satisfiable State Set of TCTL Formula ϕ.

Input: A state class graph $(R(N^t, \pi_0), E)$ and a TCTL formula ϕ;
Output: The satisfiable state set **SAT-T(ϕ)**;

1　**if** $\phi = $ **true then**
2　　　**return** $\Big\{ (\pi, C_{\tau_t}, \tau_t) \mid \pi = (M, C) \in R(N^t, \pi_0), t$ is schadulable at π, $\tau_t = \theta_{(t,\pi)}, C_{\tau_t}$
　　　　represents the clock domains of the schedulable subclass of t at $\pi \Big\}$;

3　**if** $\phi = p$ **then**
4　　　**return** $\Big\{ (\pi, C_{\tau_t}, \tau_t) \mid \pi = (M, C) \in R(N^t, \pi_0), M(p) > 0, t$ is schadulable at π,
　　　　$\tau_t = \theta_{(t,\pi)}, C_{\tau_t}$ represents the clock domains of the schedulable subclass of t at $\pi \Big\}$;

5　**if** $\phi = \neg\phi_1$ **then**
6　　　**return SAT-T(true)** $\setminus$ **SAT-T(ϕ_1)**;

7　**if** $\phi = \phi_1 \wedge \phi_2$ **then**
8　　　**return SAT-T(ϕ_1)** $\cap$ **SAT-T(ϕ_2)**;

9　**if** $\phi = E(\phi_1 \, U_I \, \phi_2)$ **then**
10　　**return SAT-T$_{EU}$(ϕ_1, ϕ_2, I)**;

11　**if** $\phi = A(\phi_1 \, U_I \, \phi_2)$ **then**
12　　**return SAT-T$_{AU}$(ϕ_1, ϕ_2, I)**;

---

the following two guided tuples are constructed according to our algorithm:

$$\Big((p_3 + p_5 + p_7 + p_9 + p_{13}, 4 \le h(t_4) \le 8 \wedge h(t_5) = 0),$$
$$4 \le h(t_4) \le 8 \wedge h(t_5) = 0, 1 \le \tau_{t_4} \le 5 \Big)$$

and

$$\Big((p_3 + p_5 + p_7 + p_9 + p_{13}, 4 \le h(t_4) \le 8 \wedge h(t_5) = 0),$$
$$4 \le h(t_4) \le 5 \wedge h(t_5) = 0, 5 \le \tau_{t_5} \le 5 \Big)$$

which are shown in the 9-th and 10-th rows of Table 9.1, respectively.

For the second guided tuple above, the guided clock domains of t_4 and t_5 are

$$4 \le h(t_4) \le 5 \wedge h(t_5) = 0,$$

i.e., the state subclass represented by this guided tuple is

$$(p_3 + p_5 + p_7 + p_9 + p_{13}, 4 \le h(t_4) \le 5 \wedge h(t_5) = 0).$$

As we know, at those states with clock domains

Table 9.1 The satisfiable states of $p_5 \wedge p_{13}$ in the state class graph in Fig 8.8, represented by guided tuples $(\pi, C_{\tau_t}, \tau_t)$

$\pi = (M, C)$	C_{τ_t}	τ_t
$\Big(p_2 + p_5 + 2p_7 + p_9 + p_{13}$ $2 \leq h(t_2) = h(t_5) \leq 5 \wedge h(t_4) = 0 \Big)$	$2 \leq h(t_2) = h(t_5) \leq$ $5 \wedge h(t_4) = 0$	$0 \leq \tau_{t_5} \leq 3$
$\Big(p_2 + p_5 + p_7 + p_{10} + p_{11} + p_{13}$ $h(t_2) = 5 \wedge h(t_7) = 0 \wedge 0 \leq h(t_4) \leq 3 \Big)$	$h(t_2) = 5 \wedge h(t_7) = 0 \wedge 0 \leq$ $h(t_4) \leq 3$	$0 \leq \tau_{t_7} \leq 0$
$\Big(p_2 + p_5 + p_7 + p_{10} + p_{11} + p_{13}$ $h(t_2) = 5 \wedge h(t_7) = 0 \wedge h(t_4) = 0 \Big)$	$h(t_2) = 5 \wedge h(t_7) =$ $0 \wedge h(t_4) = 0$	$0 \leq \tau_{t_7} \leq 0$
$\Big(p_3 + p_5 + p_7 + p_9 + p_{13}$ $h(t_4) = 0 \wedge h(t_5) = 0 \Big)$	$h(t_4) = 0 \wedge h(t_5) = 0$	$5 \leq \tau_{t_5} \leq 5$
$\Big(p_3 + p_5 + p_7 + p_9 + p_{13}$ $0 \leq h(t_4) \leq 5 \wedge h(t_5) = 0 \Big)$	$4 \leq h(t_4) \leq 5 \wedge h(t_5) = 0$	$4 \leq \tau_{t_4} \leq 5$
$\Big(p_3 + p_5 + p_7 + p_9 + p_{13}$ $0 \leq h(t_4) \leq 5 \wedge h(t_5) = 0 \Big)$	$0 \leq h(t_4) \leq 5 \wedge h(t_5) = 0$	$5 \leq \tau_{t_5} \leq 5$
$\Big(p_3 + p_5 + p_7 + p_9 + p_{13}$ $4 \leq h(t_4) \leq 5 \wedge h(t_5) = 0 \Big)$	$4 \leq h(t_4) \leq 5 \wedge h(t_5) = 0$	$4 \leq \tau_{t_4} \leq 5$
$\Big(p_3 + p_5 + p_7 + p_9 + p_{13}$ $4 \leq h(t_4) \leq 5 \wedge h(t_5) = 0 \Big)$	$4 \leq h(t_4) \leq 5 \wedge h(t_5) = 0$	$5 \leq \tau_{t_5} \leq 5$
$\Big(p_3 + p_5 + p_7 + p_9 + p_{13}$ $4 \leq h(t_4) \leq 8 \wedge h(t_5) = 0 \Big)$	$4 \leq h(t_4) \leq 8 \wedge h(t_5) = 0$	$1 \leq \tau_{t_4} \leq 5$
$\Big(p_3 + p_5 + p_7 + p_9 + p_{13}$ $4 \leq h(t_4) \leq 8 \wedge h(t_5) = 0 \Big)$	$4 \leq h(t_4) \leq 5 \wedge h(t_5) = 0$	$5 \leq \tau_{t_5} \leq 5$
$\Big(p_3 + p_5 + p_{10} + p_{11} + p_{13}$ $h(t_4) = 5 \wedge h(t_7) = 0 \Big)$	$h(t_4) = 5 \wedge h(t_7) = 0$	$0 \leq \tau_{t_7} \leq 0$
$\Big(p_3 + p_5 + p_{10} + p_{11} + p_{13}$ $5 \leq h(t_4) \leq 10 \wedge h(t_7) = 0 \Big)$	$9 \leq h(t_4) \leq 10 \wedge h(t_7) = 0$	$0 \leq \tau_{t_4} \leq 0$
$\Big(p_3 + p_5 + p_{10} + p_{11} + p_{13}$ $5 \leq h(t_4) \leq 10 \wedge h(t_7) = 0 \Big)$	$5 \leq h(t_4) \leq 10 \wedge h(t_7) = 0$	$0 \leq \tau_{t_7} \leq 0$
$\Big(p_3 + p_5 + p_{10} + p_{11} + p_{13}$ $9 \leq h(t_4) \leq 10 \wedge h(t_7) = 0 \Big)$	$9 \leq h(t_4) \leq 10 \wedge h(t_7) = 0$	$0 \leq \tau_{t_4} \leq 0$
$\Big(p_3 + p_5 + p_{10} + p_{11} + p_{13}$ $9 \leq h(t_4) \leq 10 \wedge h(t_7) = 0 \Big)$	$9 \leq h(t_4) \leq 10 \wedge h(t_7) = 0$	$0 \leq \tau_{t_7} \leq 0$

Table 9.2 The satisfiable states of $p_6 \wedge p_{13}$ in the state class graph in Fig 8.8, represented by guided tuples $(\pi, C_{\tau_t}, \tau_t)$

$\pi = (M, C)$	C_{τ_t}	τ_t
$\left(\begin{array}{l} p_3 + p_6 + p_7 + p_8 + p_9 + p_{13} \\ 4 \le h(t_5) \le 5 \end{array} \right)$	$4 \le h(t_5) \le 5$	$0 \le \tau_{t_5} \le 1$
$\left(\begin{array}{l} p_3 + p_6 + p_7 + p_8 + p_9 + p_{13} \\ 1 \le h(t_5) \le 5 \end{array} \right)$	$1 \le h(t_5) \le 5$	$0 \le \tau_{t_5} \le 4$
$\left(\begin{array}{l} p_3 + p_6 + p_8 + p_{10} + p_{11} + p_{13} \\ h(t_6) = h(t_7) = 0 \end{array} \right)$	$h(t_6) = h(t_7) = 0$	$0 \le \tau_{t_6} \le 0$
$\left(\begin{array}{l} p_3 + p_6 + p_8 + p_{10} + p_{11} + p_{13} \\ h(t_6) = h(t_7) = 0 \end{array} \right)$	$h(t_6) = h(t_7) = 0$	$0 \le \tau_{t_7} \le 0$
$\left(\begin{array}{l} p_3 + p_6 + p_9 + p_{11} + p_{12} + p_{13} \\ h(t_7) = 0 \end{array} \right)$	$h(t_7) = 0$	$0 \le \tau_{t_7} \le 0$
$\left(p_3 + p_6 + p_9 + p_{13}, - \right)$	–	–

$$5 < h(t_4) \le 8 \wedge h(t_5) = 0,$$

the transition t_5 is not schedulable because the static firing interval of t_5 is [5..5] but the static firing interval of t_4 is [9..10].

Those guided tuples standing for satisfiable states of the sub-formula $p_6 \wedge p_{13}$ are listed in Table 9.2.

> Notice that, the set-minus operation **SAT-T(true)** \ **SAT-T(ϕ)** in Line 6 of Algorithm 9.1 means the minus operation of the satisfiable state sets represented by any two guided tuples coming, respectively, from **SAT-T(true)** and **SAT-T(ϕ)**. Later, we will see such an example about the set-minus operation of two state subclasses. The set-intersection operation in Line 8 of Algorithm 9.1 is similar.

9.2.2   Computing Satisfiable State Set for EU$_I$ Based on Forward Exploration

The computation of the satisfiable state set of $\phi = E(\phi_1 \ U_I \ \phi_2)$ is shown in Algorithm 9.2. It first computes the satisfiable state sets of sub-formulae ϕ_1 and ϕ_2, denoted as X and Y, respectively.

Algorithm 9.2: SAT-T$_{\mathbf{EU}}(\phi_1, \phi_2, \mathbf{I})$: Computing the Satisfiable State Set of $\mathbf{E}(\phi_1 \ \mathbf{U_I} \ \phi_2)$.

Input: A state class graph $(R(N^t, \pi_0), E)$, two formulae ϕ_1 and ϕ_2, and an interval $\mathbf{I}$;
Output: The satisfiable state set **SAT-T$_{\mathbf{EU}}(\phi_1, \phi_2, \mathbf{I})$**;

1   $X \leftarrow$ **SAT-T**(ϕ_1);
2   $Y \leftarrow$ **SAT-T**(ϕ_2);
3   $Z \leftarrow \emptyset$;
4   **for** *each* $(\pi, C_{\tau_t}, \tau_t) \in X$ **do**
5 | Staring at π, we use the depth-first strategy to travel this state class graph to find all potentially satisfiable paths and denote them $Path(\pi, C_{\tau_t}, \tau_t)$;
6 | **for** *each path in* $Path(\pi, C_{\tau_t}, \tau_t)$ **do**
7 | | Decide if this path is reasonable based on the Alarm-clock method;
8 | | **if** *it is reasonable* **then**
9 | | | Compute the satisfiable subclass and denote it as $(\pi, C'_{\tau_t}, \tau'_t)$;
10 | | | $Z \leftarrow Z \cup \{(\pi, C'_{\tau_t}, \tau'_t)\}$;

11 **return** Z;

Then, for each guided tuple in X, we utilise the depth-first strategy to look for all possible directed paths in the state class graph, such that

1. the last node of each such path corresponds to an elements of Y,
2. other nodes correspond to the elements of X, and
3. a time interval is constructed according to the guided dynamic firing intervals of all those nodes (except the last one) in each path, and the intersection between this interval and $\mathbf{I}$ is not empty.

In other words, such a path is actually composed of the guided tuples in X and Y, but the first elements of these guided tuples of this path correspond to a directed path in the state class graph.

We assume that there are k nodes (except the last one) in such a path and their guided dynamic firing intervals are denoted as τ_{t_j} where $j \in \mathbb{N}_k^+$. Then the *constructed interval* of this path is

$$\left[\sum_{j=1}^{k} \underline{\tau_{t_j}} \ .. \ \sum_{j=1}^{k} \overline{\tau_{t_j}} \right]$$

which stands for the possible shortest and longest times running along with this path.

Obviously, for a guided tuple in X, if there is a path fulfilling the above three conditions, some (but not necessarily all) states represented by this guided tuple possibly satisfy ϕ. It is also possible that all states represented by this guided tuple do not satisfy this formula. Hence, we call such a path as *potentially satisfiable*.

Example 9.3 For the element in the 9-th row in Table 9.1, i.e., the following guided tuple:

$$\Big((p_3 + p_5 + p_7 + p_9 + p_{13}, 4 \leq h(t_4) \leq 8 \wedge h(t_5) = 0),$$
$$4 \leq h(t_4) \leq 8 \wedge h(t_5) = 0, 1 \leq \tau_{t_4} \leq 5\Big),$$

we can find the following path:

$$\Big((p_3 + p_5 + p_7 + p_9 + p_{13}, 4 \leq h(t_4) \leq 8 \wedge h(t_5) = 0),$$
$$4 \leq h(t_4) \leq 8 \wedge h(t_5) = 0, 1 \leq \tau_{t_4} \leq 5\Big)$$

$$\downarrow t_4$$

$$\Big((p_3 + p_6 + p_7 + p_8 + p_9 + p_{13}, 1 \leq h(t_5) \leq 5), 1 \leq h(t_5) \leq 5, 0 \leq \tau_{t_5} \leq 4\Big).$$

And the constructed interval of this path is [1..5] that exactly has a common part with the given interval [3.9..4.2]. Hence, this path is potentially satisfiable. The last node of this path is listed in the second row in Table 9.2.

Example 9.4 For the element in the 10-th row in Table 9.1, i.e., the following guided tuple:

$$\Big((p_3 + p_5 + p_7 + p_9 + p_{13}, 4 \leq h(t_4) \leq 8 \wedge h(t_5) = 0),$$
$$4 \leq h(t_4) \leq 5 \wedge h(t_5) = 0, 5 \leq \tau_{t_5} \leq 5\Big),$$

we may find the following two paths:

$$\Big((p_3 + p_5 + p_7 + p_9 + p_{13}, 4 \leq h(t_4) \leq 8 \wedge h(t_5) = 0),$$
$$4 \leq h(t_4) \leq 5 \wedge h(t_5) = 0, 5 \leq \tau_{t_5} \leq 5\Big)$$

$$\downarrow t_5$$

$$\Big((p_3 + p_5 + p_{10} + p_{11} + p_{13}, 9 \leq h(t_4) \leq 10 \wedge h(t_7) = 0),$$
$$9 \leq h(t_4) \leq 10 \wedge h(t_7) = 0, 0 \leq \tau_{t_4} \leq 0\Big)$$

$$\downarrow t_4$$

$$\Big((p_3 + p_6 + p_8 + p_{10} + p_{11} + p_{13}, h(t_6) = h(t_7) = 0), h(t_6) = h(t_7) = 0, 0 \leq \tau_{t_6} \leq 0\Big)$$

and

$$\Big((p_3 + p_5 + p_7 + p_9 + p_{13}, 4 \le h(t_4) \le 8 \wedge h(t_5) = 0),$$
$$4 \le h(t_4) \le 5 \wedge h(t_5) = 0, 5 \le \tau_{t_5} \le 5\Big)$$

$$\downarrow t_5$$

$$\Big((p_3 + p_5 + p_{10} + p_{11} + p_{13}, 9 \le h(t_4) \le 10 \wedge h(t_7) = 0),$$
$$9 \le h(t_4) \le 10 \wedge h(t_7) = 0, 0 \le \tau_{t_4} \le 0\Big)$$

$$\downarrow t_4$$

$$\Big((p_3 + p_6 + p_8 + p_{10} + p_{11} + p_{13}, h(t_6) = h(t_7) = 0), h(t_6) = h(t_7) = 0, 0 \le \tau_{t_7} \le 0\Big)$$

which fulfil the first two conditions. But the constructed intervals of the two paths are both $[5 + 0..5 + 0] = [5..5]$ which is disjointed with $[3.9..4.2]$. Hence, they do not fulfil our requirements.

Additionally, for this guided tuple, the following path is also explored:

$$\Big((p_3 + p_5 + p_7 + p_9 + p_{13}, 4 \le h(t_4) \le 8 \wedge h(t_5) = 0),$$
$$4 \le h(t_4) \le 5 \wedge h(t_5) = 0, 5 \le \tau_{t_5} \le 5\Big)$$

$$\downarrow t_5$$

$$\Big((p_3 + p_5 + p_{10} + p_{11} + p_{13}, 9 \le h(t_4) \le 10 \wedge h(t_7) = 0),$$
$$9 \le h(t_4) \le 10 \wedge h(t_7) = 0, 0 \le \tau_{t_7} \le 0\Big)$$

$$\downarrow t_4$$

$$\Big((p_3 + p_5 + p_{10} + p_{14}, 9 \le h(t_4) \le 10), 9 \le h(t_4) \le 10, 0 \le \tau_{t_4} \le 1\Big),$$

but the third node of this path is not in Y. Therefore, it is not a satisfiable path, either.

In a word, for this guided tuple, we cannot actually find any potentially satisfiable path, i.e., it cannot be put into the satisfiable set of

$$\mathbf{E}\Big((p_5 \wedge p_{13})\mathbf{U}_{[3.9..4.2]}(p_6 \wedge p_{13})\Big).$$

When we find a potentially satisfiable path for a guided tuple in X, e.g. the one in Example 9.3, how to compute those satisfiable states according to this guided tuple and this path? If the obtained interval is a subset of $\mathbf{I}$, then nothing is changed, i.e., all states represented by this guided tuple are satisfiable. Otherwise, we take the *Alarm-clock* method [1] to decide if this path is really a satisfiable one and then to compute a new guided tuple (i.e., a subclass of the original guided tuple) satisfying this formula (if this path is really satisfiable).

- First, a virtual transition, say t_0, is set up and $\mathbf{I}$ in ϕ as its static firing interval. Along with this path, we compute the changes of the clock domain of t_0 but we do not schedule it to fire.
- If the clock domain of t_0 is still reasonable at the last node of this path, this path is really satisfiable and then we go backward to shorten the dynamic firing intervals of other transitions until the first node. At this time, we obtain this new guided tuple.
- If t_0 must be fired (i.e. the lower bound of its dynamic firing time is equal to its latest firing time) before reaching the last node, this path is not a satisfiable one and thus should be discarded.
- Specifically speaking, at the first node of this path, the clock of t_0 begins to work, i.e., $h(t_0) = 0$; then, according to the static firing interval of t_0, we re-compute the guided dynamic firing interval and the guided clock domains of a node along with the path, until the last node.
- When we go backward to compute this new guided tuple, we utilise those new guided dynamic firing intervals and new guided clock domains to re-compute the subclass of each predecessor until the first node. For this computation, Algorithm 8.4 in Chap. 8 is used.

Next, we use two instances to illustrate the above computations.

Example 9.5   We consider the path computed in Example 9.3. The static firing interval of t_0 is [3.9..4.2]. We add the clock domain of t_0 into the first node as follows:

$$\Big((p_3 + p_5 + p_7 + p_9 + p_{13}, 4 \le h(t_4) \le 8 \wedge h(t_5) = 0),$$
$$4 \le h(t_4) \le 8 \wedge h(t_5) = 0 \wedge h(t_0) = 0, 1 \le \tau_{t_4} \le 4.2\Big).$$

In the original guided tuple, the guided dynamic firing interval of t_4 is [1..5], but now it should be [1..4.2] due to the constraint of the static firing interval of t_0. Hence, we obtain its next node (i.e., the last node) as follows

$$\Big((p_3 + p_6 + p_7 + p_8 + p_9 + p_{13}, 1 \le h(t_5) \le 5), 1 \le h(t_5) \le 4.2 \wedge 1 \le h(t_0) \le 4.2, --\Big).$$

Since the clock domain of t_0 is reasonable in the last node, this path is a satisfiable one. According to the new dynamic firing interval [1..4.2] of t_4, we utilise Algorithm 8.4 in Chap. 8 to compute a subclass of the first node as follows:

$$(p_3 + p_5 + p_7 + p_9 + p_{13}, 4.8 \le h(t_4) \le 8 \wedge h(t_5) = 0).$$

On the other hand, all states represented by the complement of the above subclass (relative to this guided tuple) is

$$(p_3 + p_5 + p_7 + p_9 + p_{13}, 4 \leq h(t_4) < 4.8 \wedge h(t_5) = 0),$$

that does not satisfy the formula

$$\mathbf{E}\Big((p_5 \wedge p_{13})\mathbf{U}_{[3.9..4.2]}(p_6 \wedge p_{13})\Big).$$

Therefore, we obtain the following new guided tuple satisfying the above formula:

$$\Big((p_3 + p_5 + p_7 + p_9 + p_{13}, 4 \leq h(t_4) \leq 8 \wedge h(t_5) = 0),$$
$$4.8 \leq h(t_4) \leq 8 \wedge h(t_5) = 0, 1 \leq \tau_{t_4} \leq 4.2\Big),$$

whereas the original guided tuple is

$$\Big((p_3 + p_5 + p_7 + p_9 + p_{13}, 4 \leq h(t_4) \leq 8 \wedge h(t_5) = 0),$$
$$4 \leq h(t_4) \leq 8 \wedge h(t_5) = 0, 1 \leq \tau_{t_4} \leq 5\Big).$$

Example 9.6 We consider another instance of a longer path in Fig 8.8. The path is

$$\Big((p_2 + p_5 + p_7 + p_{10} + p_{14}, h(t_2) = 5 \wedge 0 \leq h(t_4) \leq 3),$$
$$h(t_2) = 5 \wedge 0 \leq h(t_4) \leq 3, 4 \leq \tau_{t_2} \leq 5\Big)$$

$$\downarrow t_2$$

$$\Big((p_3 + p_5 + p_7 + p_8 + p_{10} + p_{14}, 4 \leq h(t_4) \leq 8 \wedge h(t_6) = 0),$$
$$4 \leq h(t_4) \leq 8 \wedge h(t_6) = 0, 0 \leq \tau_{t_6} \leq 0\Big)$$

$$\downarrow t_6$$

$$\Big((p_3 + p_5 + p_7 + p_9 + p_{12} + p_{14}, 4 \leq h(t_4) \leq 8 \wedge h(t_5) = h(t_8) = 0),$$
$$4 \leq h(t_4) \leq 8 \wedge h(t_5) = h(t_8) = 0, 0 \leq \tau_{t_8} \leq 0\Big)$$

$$\downarrow t_8$$

$$\Big((p_3 + p_5 + p_7 + p_9 + p_{13}, 4 \leq h(t_4) \leq 8 \wedge h(t_5) = 0),$$
$$4 \leq h(t_4) \leq 8 \wedge h(t_5) = 0, 1 \leq \tau_{t_4} \leq 5\Big)$$

$$\downarrow t_4$$

$$\Big((p_3 + p_6 + p_7 + p_8 + p_9 + p_{13}, 1 \le h(t_5) \le 5), 1 \le h(t_5) \le 5, 1 \le \tau_{t_5} \le 4\Big).$$

The time interval of this path is

$$[5..10] = [4+0+0+1 .. 5+0+0+5].$$

Case 1: If the time interval **I** of some **EU**$_\mathbf{I}$ formula is $[4..6]$, then this path is a potentially satisfiable one due to

$$[4..6] \cap [5..10] \neq \emptyset.$$

Then, we set up a virtual transition t_0 with the static firing interval $[4..6]$ and let it take part in this running. We obtain the following path:

$$\Big((p_2 + p_5 + p_7 + p_{10} + p_{14}, h(t_2) = 5 \wedge 0 \le h(t_4) \le 3),$$
$$h(t_2) = 5 \wedge 0 \le h(t_4) \le 3 \wedge h(t_0) = 0, 4 \le \tau_{t_2} \le 5\Big)$$

$$\downarrow t_2$$

$$\Big((p_3 + p_5 + p_7 + p_8 + p_{10} + p_{14}, 4 \le h(t_4) \le 8 \wedge h(t_6) = 0),$$
$$4 \le h(t_4) \le 8 \wedge h(t_6) = 0 \wedge 4 \le h(t_0) \le 5 \wedge 0 \le h(t_4) - h(t_0) \le 3, 0 \le \tau_{t_6} \le 0\Big)$$

$$\downarrow t_6$$

$$\Big((p_3 + p_5 + p_7 + p_9 + p_{12} + p_{14}, 4 \le h(t_4) \le 8 \wedge h(t_5) = h(t_8) = 0),$$
$$4 \le h(t_4) \le 8 \wedge h(t_5) = h(t_8) =$$
$$0 \wedge 4 \le h(t_0) \le 5 \wedge 0 \le h(t_4) - h(t_0) \le 3, 0 \le \tau_{t_8} \le 0\Big)$$

$$\downarrow t_8$$

$$\Big((p_3 + p_5 + p_7 + p_9 + p_{13}, 4 \le h(t_4) \le 8 \wedge h(t_5) = 0),$$
$$7 \le h(t_4) \le 8 \wedge h(t_5) = 0 \wedge 4 \le h(t_0) \le 5 \wedge 0 \le h(t_4) - h(t_0) \le 3, 1 \le \tau_{t_4} \le 2\Big)$$

$$\downarrow t_4$$

$$\Big((p_3 + p_6 + p_7 + p_8 + p_9 + p_{13}, 1 \le h(t_5) \le 5), 1 \le h(t_5) \le 2 \wedge 5 \le h(t_0) \le 6, --\Big).$$

Since the clock domain of t_0 in the last node is still reasonable, i.e.,

$$[5..6] \subseteq [4..6],$$

it is a satisfiable computation path.

Consequently, we backward compute the subclasses of all nodes as follows:

$$\Big((p_2 + p_5 + p_7 + p_{10} + p_{14}, h(t_2) = 5 \wedge 0 \le h(t_4) \le 3),$$
$$h(t_2) = 5 \wedge h(t_4) = 3 \wedge h(t_0) = 0, 4 \le \tau_{t_2} \le 5\Big)$$

$$\uparrow \downarrow t_2$$

$$\Big((p_3 + p_5 + p_7 + p_8 + p_{10} + p_{14}, 4 \le h(t_4) \le 8 \wedge h(t_6) = 0),$$
$$7 \le h(t_4) \le 8 \wedge h(t_6) = 0 \wedge 4 \le h(t_0) \le 5 \wedge 0 \le h(t_4) - h(t_0) \le 3, 0 \le \tau_{t_6} \le 0\Big)$$

$$\uparrow \downarrow t_6$$

$$\Big((p_3 + p_5 + p_7 + p_9 + p_{12} + p_{14}, 4 \le h(t_4) \le 8 \wedge h(t_5) = h(t_8) = 0),$$
$$7 \le h(t_4) \le 8 \wedge h(t_5) = h(t_8) =$$
$$0 \wedge 4 \le h(t_0) \le 5 \wedge 0 \le h(t_4) - h(t_0) \le 3, 0 \le \tau_{t_8} \le 0\Big)$$

$$\uparrow \downarrow t_8$$

$$\Big((p_3 + p_5 + p_7 + p_9 + p_{13}, 4 \le h(t_4) \le 8 \wedge h(t_5) = 0),$$
$$7 \le h(t_4) \le 8 \wedge h(t_5) = 0 \wedge 4 \le h(t_0) \le 5 \wedge 0 \le h(t_4) - h(t_0) \le 3, 1 \le \tau_{t_4} \le 2\Big)$$

$$\uparrow \downarrow t_4$$

$$\Big((p_3 + p_6 + p_7 + p_8 + p_9 + p_{13}, 1 \le h(t_5) \le 5), 1 \le h(t_5) \le 2 \wedge 5 \le h(t_0) \le 6, --\Big).$$

As a result, we obtain a new guided tuple as follows:

$$\Big((p_2 + p_5 + p_7 + p_{10} + p_{14}, h(t_2) = 5 \wedge 0 \le h(t_4) \le 3),$$
$$h(t_2) = 5 \wedge h(t_4) = 3, 4 \le \tau_{t_2} \le 5\Big).$$

Case 2: If the time interval **I** of some EU_I formula is $[5..5.1]$, then this path is a potentially satisfiable one, too, due to

$$[5..5.1] \cap [5..10] \ne \emptyset.$$

Then, we set up a virtual transition t_0 with the static firing interval [5..5.1] and let it take part in this running. We obtain the following computing procedure:

$$\Big((p_2 + p_5 + p_7 + p_{10} + p_{14}, h(t_2) = 5 \wedge 0 \le h(t_4) \le 3),$$

$$h(t_2) = 5 \wedge 0 \le h(t_4) \le 3 \wedge h(t_0) = 0, 4 \le \tau_{t_2} \le 5\Big)$$

$$\downarrow t_2$$

$$\Big((p_3 + p_5 + p_7 + p_8 + p_{10} + p_{14}, 4 \le h(t_4) \le 8 \wedge h(t_6) = 0),$$

$$4 \le h(t_4) \le 8 \wedge h(t_6) = 0 \wedge 4 \le h(t_0) \le 5 \wedge 0 \le h(t_4) - h(t_0) \le 3, 0 \le \tau_{t_6} \le 0\Big)$$

$$\downarrow t_6$$

$$\Big((p_3 + p_5 + p_7 + p_9 + p_{12} + p_{14}, 4 \le h(t_4) \le 8 \wedge h(t_5) = h(t_8) = 0),$$

$$4 \le h(t_4) \le 8 \wedge h(t_5) = h(t_8) =$$

$$0 \wedge 4 \le h(t_0) \le 5 \wedge 0 \le h(t_4) - h(t_0) \le 3, 0 \le \tau_{t_8} \le 0\Big)$$

$$\downarrow t_8$$

$$\Big((p_3 + p_5 + p_7 + p_9 + p_{13}, 4 \le h(t_4) \le 8 \wedge h(t_5) = 0),$$

$$4 \le h(t_4) \le 8 \wedge h(t_5) = 0 \wedge 4 \le h(t_0) \le 5 \wedge 0 \le h(t_4) - h(t_0) \le 3, ? \le \tau_{t_4} \le ?\Big)$$

However, when we produce the above node, we can see that t_4 is not schedulable any more, but it should be scheduled to fire in the original path. Why $\tau_{t_4} = \emptyset$? Because the static firing interval of t_4 is [9..10] and $4 \le h(t_4) \le 8$, we know that the dynamic firing interval of t_4 should be looked for in [1..6]. Additionally, because the static firing interval of t_0 is [5..5.1] and $4 \le h(t_4) \le 5$, we know that the maximum value of the dynamic firing interval of t_4 should not be beyond 1.1. Consequently, the dynamic firing interval of t_4 should be looked for in [1..1.1]. At this time, if we want to schedule t_4, then its clock domain should be

$$7.9 \le h(t_4) \le 8.$$

On the other hand, because the latest firing time of t_0 is 5.1, we know that the clock domain of t_4 should be

$$4 \le h(t_0) \le 4.1.$$

However, $h(t_4)$ and $h(t_0)$ must satisfy

$$0 \le h(t_4) - h(t_0) \le 3.$$

Obviously, there is no any clock value of t_4 satisfying the above three conditions, i.e., t_4 is not schedulable. Therefore, this path is not satisfiable for an $\mathbf{EU_I}$ formula when $\mathbf{I} = [5..5.1]$.

9.2.3 Computing Satisfiable State Set of $\mathbf{AU_I}$ Based on Forward Exploration

The computation of the satisfiable state set of a TCTL formula of $\mathbf{AU_I}$ is more complex and difficult. From Fig. 8.8 we can see that for different state classes with the same marking, their clock domains possibly have some common parts. And from Table 9.1 we also see that for some different guided tuples with the same state class, their guided clock domains also have some common parts. In other words, different guided tuples possibly have some common states. In consequence, it is possible that some state satisfies a formula of $\mathbf{AU_I}$ at some guided tuple according to all paths starting at this guided tuple, but it does not satisfy this formula at another guided tuple according to all paths starting at this one; in other words, such a state actually does not satisfy this formula.

To compute the satisfiable state set of $\mathbf{A}(\phi_1 \ \mathbf{U_I} \ \phi_2)$, we should solve the following two problems:

1. Problem 1: Given a guided tuple in $\mathbf{SAT\text{-}T}(\phi_1)$, how to compute the *potentially satisfiable* state set of $\mathbf{A}(\phi_1 \ \mathbf{U_I} \ \phi_2)$ at this guided tuple according to all paths starting at this guided tuple?
2. Problem 2: Based on the *potentially satisfiable* sets computed at these different guided tuples, how to compute those *really satisfiable* states?

Notice that even when all paths, starting at a state of a given guided tuple, satisfy an $\mathbf{AU_I}$ formula, this does not necessarily means all paths, starting at this state in the whole state class graph, satisfy it, because these paths computed according to the given guided tuple comes from the field constrained by this guided tuple and thus cannot stand for a whole. Therefore, such a state is just a *potential satisfiable* one. Later, we will see such an example.

For Problem 1, we still use the depth-first strategy to travel a state class graph to explore all possible paths starting at a guided tuple where this guided tuple is an element in $X = \mathbf{SAT\text{-}T}(\phi_1)$. If there is an unsatisfiable path for this guided tuple, then all states represented by this guided tuple does not satisfy $\mathbf{A}(\phi_1 \ \mathbf{U_I} \ \phi_2)$, and this guided tuple is labelled as *unsatisfiable*.

The case, that all paths starting at a guided tuple are satisfiable, does not imply all states represented by this guided tuple are satisfiable, as shown in Examples 9.5 and 9.6. For this case, we should first compute the satisfiable subclasses corresponding to these paths. Then, all states in their intersection satisfy this formula at this guided tuple, and the remainders do not satisfy it. Hence, the former is labelled as *potentially satisfiable* and the latter as *unsatisfiable*.

For Problem 2, we first divide the guided tuples in $X = \textbf{SAT-T}(\phi_1)$ into different groups. If two guided tuples $(\pi_1, C_{\tau_{t_1}}, \tau_{t_1})$ and $(\pi_2, C_{\tau_{t_2}}, \tau_{t_2})$ fulfil $\pi_1 = \pi_2$, then they are put into the same group. After we obtain all new guided tuples for all guided tuples in a group and label them by *unsatisfiable* or *potentially satisfiable*, we remove from the *potentially satisfiable* guided tuples those states labelled by *unsatisfiable* and finally obtain *really satisfiable* states. Algorithm 9.3 describes this procedure.

We consider the following formula

$$\textbf{A}\Big((p_5 \wedge p_{13})\textbf{U}_{[3.9..4.2]}(p_6 \wedge p_{13})\Big)$$

w.r.t. the time Petri net in Fig. 8.7a and the state class graph in Fig. 8.8. Those guided tuples in $X = \textbf{SAT-T}(p_5 \wedge p_{13})$ and $Y = \textbf{SAT-T}(p_6 \wedge p_{13})$ have been listed in Tables 9.1 and 9.2, respectively. Next, we consider the two guided tuples in the 9-th and 10-th rows in Table 9.1, because they corresponds to the same state class.

Example 9.7 For the guided tuple in the 9-th row of Table 9.1, i.e.,

$$\Big((p_3 + p_5 + p_7 + p_9 + p_{13}, 4 \le h(t_4) \le 8 \wedge h(t_5) = 0),$$
$$4 \le h(t_4) \le 8 \wedge h(t_5) = 0, 1 \le \tau_{t_4} \le 5\Big),$$

we have shown in Examples 9.3 and 9.5 that there is a unique satisfiable path, but only the state subclass

$$(p_3 + p_5 + p_7 + p_9 + p_{13}, 4.8 \le h(t_4) \le 8 \wedge h(t_5) = 0)$$

is satisfiable for the above formula according to all paths starting at the above guided tuple. Hence, these states are labelled by *potentially satisfiable*.

On the other hand, the complement of these potentially satisfiable states relative to the above guided tuple are labelled by *unsatisfiable*, i.e.,

$$(p_3 + p_5 + p_7 + p_9 + p_{13}, 4 \le h(t_4) < 4.8 \wedge h(t_5) = 0).$$

Algorithm 9.3: SAT-T$_{AU}$(ϕ_1, ϕ_2, I): Computing the Satisfiable State Set of
A(ϕ_1 U$_I$ ϕ_2).

Input: A state class graph $(R(N^t, \pi_0), E)$, two formulae ϕ_1 and ϕ_2, and an interval **I**;
Output: The satisfiable state set **SAT-T$_{AU}$**(ϕ_1, ϕ_2, **I**);
1 $X \leftarrow$ **SAT-T**(ϕ_1);
2 $Y \leftarrow$ **SAT-T**(ϕ_2);
3 $Z \leftarrow \emptyset$;
4 Divide X into $X_1, \cdots$, and X_k according to the same state class;
5 **for** *each X_j with π_j where $j \in \mathbb{N}_k^+$* **do**
6 $\quad$ *potentially-satisfiable* $\leftarrow \emptyset$;
7 $\quad$ *unsatisfiable* $\leftarrow \emptyset$;
8 $\quad$ **for** *each $(\pi_j, C_{\tau_t}, \tau_t) \in X_j$* **do**
9 $\quad\quad$ Compute the satisfiable clock domain C'_{τ_t} and the unsatisfiable clock domain

$$C''_{\tau_t} = C_{\tau_t} - C'_{\tau_t};$$

$\quad\quad$ *potentially-satisfiable* $\leftarrow$ *potentially-satisfiable* $\cup$ C'_{τ_t};
10 $\quad\quad$ *unsatisfiable* $\leftarrow$ *unsatisfiable* $\cup$ C''_{τ_t};
11 $\quad$ Compute the new guided clock domain *potentially-satisfiable* $-$ *unsatisfiable* as well as
$\quad$ the guided dynamic firing intervals;
12 $\quad$ Put new guided tuples into Z;
13 **return** Z;

In Example 9.4 we have shown that there is no satisfiable path for the guided tuple
in the 10-th row of Table 9.1, where this guided tuple is

$$\Big((p_3 + p_5 + p_7 + p_9 + p_{13}, 4 \leq h(t_4) \leq 8 \wedge h(t_5) = 0),$$
$$4 \leq h(t_4) \leq 5 \wedge h(t_5) = 0, 5 \leq \tau_{t_5} \leq 5 \Big).$$

Therefore, all states represented by this guided tuple is labelled by *unsatisfiable*
for the above formula, i.e.,

$$(p_3 + p_5 + p_7 + p_9 + p_{13}, 4 \leq h(t_4) \leq 5 \wedge h(t_5) = 0).$$

Because only the above two guided tuples correspond to the same state class, we
obtain those *really satisfiable* states computed by the following operation

really-satisfiable $=$ *potentially-satifiable* $-$ *unsatisfiable*

$$= (p_3 + p_5 + p_7 + p_9 + p_{13}, 4.8 \leq h(t_4) \leq 8 \wedge h(t_5) = 0)$$
$$- \Big((p_3 + p_5 + p_7 + p_9 + p_{13}, 4 \leq h(t_4) < 4.8 \wedge h(t_5) = 0)$$

$$\cup (p_3 + p_5 + p_7 + p_9 + p_{13}, 4 \le h(t_4) \le 5 \wedge h(t_5) = 0) \big)$$

$$= \quad (p_3 + p_5 + p_7 + p_9 + p_{13}, 5 < h(t_4) \le 8 \wedge h(t_5) = 0).$$

As a result, a new guided tuple

$$\Big((p_3 + p_5 + p_7 + p_9 + p_{13}, 4 \le h(t_4) \le 8 \wedge h(t_5) = 0),$$

$$5 < h(t_4) \le 8 \wedge h(t_5) = 0, 1 \le \tau_{t_4} \le 5 \Big)$$

is generated, that really satisfies

$$\mathbf{A}\Big((p_5 \wedge p_{13}) \mathbf{U}_{[3.9..4.2]} (p_6 \wedge p_{13}) \Big).$$

Similarly, we can carry out the above computation for other guided tuples that are listed in Table 9.1 and correspond to the same state class. The computation results illustrate that there is no other states satisfying the above formula.

9.3　Summery and Further Reading

This chapter introduces some basic knowledge about TCTL and proposes a model checking method over the state class graphs of time Petri nets. In theory, the satisfiability problem of TCTL is undecidable [2], but for bounded time Petri net, it is decidable and PSPACE-complete [3]. In addition to the checking method based on state classes, another commonly-used one is based on regions [2–5]. In [6], there is a systematic description of region-based TCTL model checking. Several mature tools that can check TCTL are UPPAAL [7, 8], Romeo [9], and PAT [10]. Some applications of Petri nets and TCTL can be found in [11–15].

References

1. Hadjidj, R., Boucheneb, H.: (2009). On-the-fly TCTL model checking for time Petri nets. *Theoretical Computer Science* 410: 4241–4261 (2009)
2. Alur, R., Courcoubetis, C., Dill, D. Model-checking for real-time systems. In: *the Fifth Annual IEEE Symposium on Logic in Computer Science*, pp. 414–425 (1990)
3. Boucheneb, H., Gardey, G., Roux, O. H.: TCTL model checking of time Petri nets. *Journal of Logic and Computation* 19: 1509–1540 (2009)
4. Larsen, K.G., Pettersson, P., Wang, Y.: Model-checking for real-time systems. In: *International Symposium on Fundamentals of Computation Theory)*, pp. 62–88 (1995)
5. Sun, J., Liu, Y., Dong, J.S., Liu, Y., Shi, L., André,É.: Modeling and verifying hierarchical real-time systems using stateful timed CSP. *ACM Transactions on Software Engineering and Methodology* 22: 1–29. (2013)
6. Baier, C., Katoen, J.P.: *Principles of Model Checking*. MIT Press (2008)

7. Bengtsson, J., Larsen, K., Larsson, F., Pettersson, P., Wang, Y.: UPPAAL – a tool suite for automatic verification of real-time systems. In: *International Hybrid Systems Workshop*, pp. 232–243 (1995)
8. Behrmann, G., David, A., Larsen, K.G.: A tutorial on UPPAAL. *Formal methods for the design of real-time systems*, 200–236 (2004)
9. Lime, D., Roux, O.H., Seidner, C., Traonouez, L.M.: Romeo: A parametric model-checker for Petri nets with stopwatches. In: *International Conference on Tools and Algorithms for the Construction and Analysis of Systems*, pp. 54–57 (2009)
10. Sun, J., Liu, Y., Dong, J.S., Pang, J.: PAT: Towards flexible verification under fairness. In: *International conference on computer aided verification*, pp. 709–714 (2009)
11. He, L.F., Liu, G.J.: Time-point-interval prioritized time Petri nets modelling real-time systems and TCTL checking. *Journal of Software* 33: 2947–2963 (2022) (in Chinese, https://doi.org/10.13328/j.cnki.jos.006607)
12. Jbeli, N., Sbai, Z.: On improving model checking of time Petri nets and its application to the formal verification. *International Journal of Service Science, Management, Engineering, and Technology* 12: 68–84 (2021)
13. Parrot, R., Briday, M., Roux, O.H.: Timed Petri nets with reset for pipelined synchronous circuit design. In: *2021 International Conference on Applications and Theory of Petri Nets and Concurrency*, pp. 55–75 (2021)
14. Coullon, H., Jard, C., Lime, D.: Integrated model-checking for the design of safe and efficient distributed software commissioning. In: *International Conference on Integrated Formal Methods*, pp. 120–137 (2019)
15. Haur, I., Béchennec, J.L., Roux, O.H.: Formal schedulability analysis based on multi-core RTOS model. In: *the 29th International Conference on Real-Time Networks and Systems* pp. 216-225 (2021)

Chapter 10
Plain Time Petri Nets with Priorities

10.1 Plain Time Petri Net with Priorities

We augment priorities into a time Petri net, named *plain time Petri nets with priorities*. The aim is to model and analysis some real-time system about scheduling tasks. This section introduces the definition and firing rules of plain time Petri nets with priorities as well as their state class graphs. For some similar work, one may read [1].

10.1.1 Definition of Plain Time Petri Net with Priorities

Definition 10.1 (*Plain Time Petri Net With Priorities*) A *plain time Petri net with priorities* is a 7-tuple $(N^p, M_0) \triangleq (P, T_S, T_U, F, \mathcal{I}, \mathcal{P}, M_0)$ where

1. $(P, T_S \cup T_U, F, \mathcal{I}, M_0)$ is a time Petri net such that for each $t \in T_S \cup T_U$:

$$\underline{\mathcal{I}(t)} = \overline{\mathcal{I}(t)},$$

2. T_S is a set of *suspendible transitions*, T_U is a set of *un-suspendible transitions*, $T_S \cap T_U = \emptyset$, and
3. $\mathcal{P}: T_S \cup T_U \to \mathbb{N}$ is a *priority function*.

We call it *plain* just because the earliest and latest firing times of each transition are identical. Under this condition, ∞ does not occur in the static firing interval of any transition.

Transitions are partitioned into two classes: *suspendible* and *un-suspendible*. As we know, when a schedulable transition becomes disabled or newly enabled in a time Petri net, its clock is reset by 0. Here, if such a transition is suspendible, then its clock will stop but the current time is recorded; once it becomes schedulable again (no matter how much time has been gone), this clock will go on working from this recorded time point. When two transitions may both

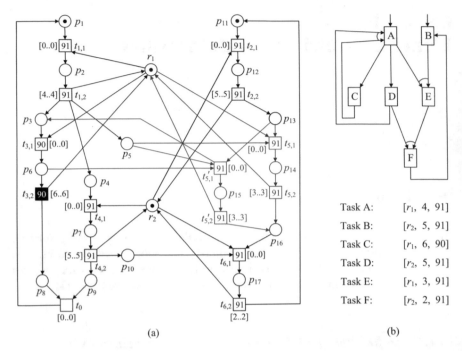

Fig. 10.1 **a** A plain time Petri net with priorities, and **b** its corresponding task dependence diagram as well as information of all tasks

be fired according to the firing rules of time Petri nets, only the one with a higher priority will be scheduled to fire; certainly, if firing this transition disables other suspendible transitions (especially, their clocks have worked), then they will be suspended and their clocks' values are recorded.

Before formalising these firing rules, we first show a plain time Petri net with priorities as shown in Fig. 10.1a.

In this net diagram, the values of priorities are written in all transitions, and a solid box represents a suspendible transition (e.g. $t_{3,2}$). In fact, it models the real-time system in Fig. 10.1b. In this system, there are 6 tasks with different priorities and running times. And they are performed by two machines r_1 and r_2.

Since Tasks C and E share r_1 and the latter has a higher priority, r_1 is possibly preempted by the latter when it is being used by the former. Although Task A has a higher priority than Task C and they also share r_1, the case that the former preempts the latter's resource cannot take place because they have a directly-dependent relation on this control flow. In consequence, we use transitions $t_{5,1}$ and $t_{5,2}$ to simulate Task E obtaining and releasing r_1 normally, but we use transitions $t'_{5,1}$ and $t'_{5,2}$ to simulate Task E preempting and releasing r_1.

When Task E is driven (i.e., places p_5 and p_{13} both have a token, representing that Tasks A and B have finished.) and the machine r_1 is idle, $t_{5,1}$ rather than $t'_{5,1}$ can be fired, standing for a normal allocation of r_1. If Task E is driven but r_1 is occupied by Task C at this time, then r_1 is preempted by Task E, i.e., $t'_{5,1}$ rather than $t_{5,1}$ can be fired. Therefore, the place p_6 is an input of $t'_{5,1}$. The purpose that the place p_3 is an output of $t'_{5,1}$ is to represent that Task C will take part in the application for r_1 again. Certainly, the token in p_6 is removed by $t'_{5,1}$, the clock of the suspended transition $t_{3,2}$ will stop working but its current time is recorded.

Here, we use $t'_{5,1}$, p_{15}, and $t'_{5,2}$ to model, respectively, that the resource r_1 is preempted, Task E is performed, and this resource is released. In fact, the place p_{15} (respectively the transition $t'_{5,2}$) can be integrated with the place p_{14} (respectively the transition $t_{5,2}$), because they both represent performing this task (respectively releasing this resource). But for clearly illustrating their different progresses, we take such a relatively complex structure.

For a task in such a system, we usually use a pair of transitions to simulate the scheduling and terminating of this task. For instance of Task A, a token in the place p_1 means that this task is driven. Firing the transition $t_{1,1}$ means that this task is scheduled to be performed by r_1. The token in the place p_2 means that this task is being performed, and thus firing the transition $t_{1,2}$ means that this task is finished and the resource r_1 is released. Since r_1 spends 4 units of time to perform Task A, we set the static firing intervals of $t_{1,1}$ and $t_{1,2}$ are [0..0] and [4..4], respectively. The interval [0..0] of $t_{1,1}$ means that Task A is immediately scheduled once it is driven and r_1 is idle. If we want to represent such a case that it will spend some time (say 2 units of time) to schedule this task onto this machine (when they are both ready), we can set up the static firing interval of $t_{1,1}$ by [2..2], but the static firing interval of $t_{1,2}$ should be set up by [6..6] accordingly.

10.1.2   Firing Rules and State Class Graph of Plain Time Petri Nets with Priorities

We also denote $s_0 = (M_0, h_0)$ as the *initial state* of a plain time Petri net with priorities where M_0 is the initial marking and $h_0(t) = 0$ for all $t \in En(M_0)$. Notice, $En(M)$ and $NEn(M, t)$ are the same as those of time Petri nets. But, the definition of a *state* of a plain time Petri net with priorities considers the clocks of those *suspended* transitions besides enabled ones.

To this end, we use $Su(s)$ to denote the set of those suspended transitions at a state s. It is dynamically changed and will be computed in the procedure of computing a new state. Obviously, $Su(s_0) = \emptyset$.

Definition 10.2 (*State of Plain Time Petri Net with Priorities*) A *state* of a plain time Petri net with priorities is denoted as a couple $s = (M, h)$ where M is a marking and $h: En(M) \cup Su(s) \to \mathbb{R}$ is a *clock function* such that for each $t \in En(M) \cup Su(s)$:

$$h(t) \leq \overline{I(t)}.$$

Definition 10.3 (*State Transfer of Plain Time Petri Net with Priorities*) Let $s = (M, h)$ and $s' = (M', h')$ be two states of a plain time Petri net with priorities. Then, there exist the following two rules of state transfer:

1. $s \xrightarrow{\tau} s'$ if and only if s' is reachable from s by the elapsing of time $\tau \in \mathbb{R}$, i.e.,

$$\begin{cases} M' = M \\ \forall t \in En(M'): h'(t) = h(t) + \tau \leq \overline{I(t)} \\ \forall t \in Su(s): h'(t) = h(t) \\ Su(s') = Su(s) \end{cases}.$$

2. $s \xrightarrow{t} s'$ if and only if s' is immediately reachable from s by the firing of the transition $t \in En(M)$, i.e.,

$$\begin{cases} M[t\rangle M' \\ \underline{I(t)} = h(t) = \overline{I(t)} \\ \forall t' \in En(M) \setminus \{t\}: \underline{I(t')} = h(t') = \overline{I(t')} \Rightarrow \mathcal{P}(t') \leq \mathcal{P}(t) \\ \forall t' \in T_S \cap \left(En(M) \setminus En(M') \right): t' \in Su(s') \wedge h'(t') = h(t') \\ \forall t' \in Su(s): t' \notin En(M') \Rightarrow \left(t' \in Su(s') \wedge h'(t') = h(t') \right) \\ \forall t' \in Su(s) \cap En(M'): t' \notin Su(s') \wedge h'(t') = h(t') \\ \forall t' \in NEn(M', t) \setminus Su(s): h'(t') = 0 \\ \forall t' \in En(M') \setminus NEn(M', t): h'(t') = h(t') \end{cases}.$$

In the first rule, the clock value of every enabled transition pluses τ, the clock value of every being-suspended transition keeps unchanged, and the set of being-suspended transitions also keeps unchanged.

The following explains the computation of suspended transitions and the modification of the clock values at a new state generated by the second rule.

- $t' \in T_S \cap \left(En(M) \setminus En(M') \right)$ means that this suspendible transition was enabled at s but becomes disabled at s' (due to the firing of t); hence it becomes a suspended transition and its clock value will keep unchanged until it becomes enabled again.

- $t' \in Su(s) \wedge t' \notin En(M')$ means that this transition was suspended at s and now is still suspended at s' (since it is disabled either at M'); hence, it belongs to $Su(s')$ and its clock value keeps unchanged.

- $t' \in Su(s) \cap En(M')$ means that it was suspended at s but now it becomes enabled at s'; hence, it is not a suspended one any more at this new state, and its clock begins to work at the recorded value.

- $t' \in NEn(M', t) \setminus Su(s)$ means that it is newly enabled but does not come from the suspended set; hence, its clock begins to work at the value 0. This point and the above one exactly consider the two cases of newly enabled transitions: the above one considers those newly enabled transitions that are waked up from the suspended states, while this one considers those purely newly enabled transitions.

Example 10.1 Figure 10.2 is temporally viewed as a solo state graph of the plain time Petri net with priorities in Fig. 10.1a. Each edge labels the needed time of firing the related transition. For example, the state

$$(p_4 + p_5 + p_6 + p_{13} + r_2, h(t_{3,2}) = 1 \wedge h(t_{4,1}) = h(t'_{5,1}) = 0)$$

means that Task C has been performed for 1 unit of time and the resource r_2 is idle. At this state, Tasks D and E are ready. Since Task D needs r_2 and r_2 is available, it can be scheduled. Task E needs r_1 but r_1 is occupied by Task C. However, since Task E has a higher priority than Task C, it can preempt the resource r_1 from Task C, i.e., firing $t'_{5,1}$. Firing it leads to the state

$$(p_3 + p_4 + p_{15} + r_2, h(t_{3,2}) = 1 \wedge h(t_{4,1}) = h(t'_{5,2}) = 0)$$

in which the clock of the suspended transition $t_{3,2}$ stops to work but maintains the current value (implying that Task C is suspended). At this state, the transition $t_{4,1}$ can and must be scheduled and then the following state is reached:

$$(p_3 + p_7 + p_{15}, h(t_{3,2}) = 1 \wedge h(t_{4,2}) = h(t'_{5,2}) = 0).$$

This state means that Tasks D and E are performed on r_2 and r_1, respectively, while Task C is suspended. Since Task E needs 3 units of time shorter than the time needed by Task D, only the transition $t'_{5,2}$ can and must be fired after 3 units of time. In consequence, the following state is reached:

$$(p_3 + p_7 + p_{16} + r_1, h(t_{3,2}) = 1 \wedge h(t_{4,2}) = 3 \wedge h(t_{3,1}) = 0).$$

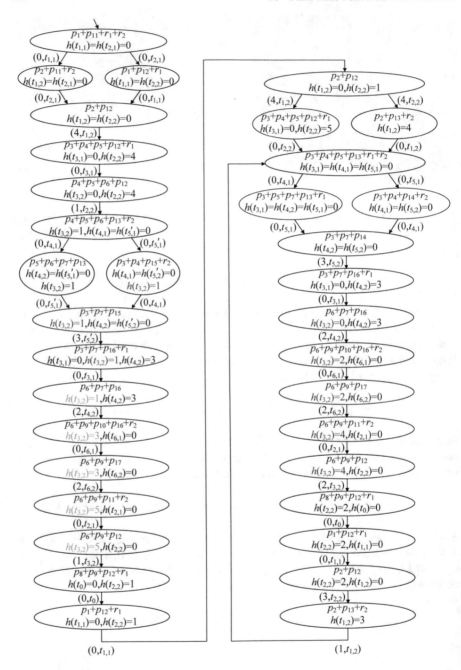

Fig. 10.2 The state class graph of the plain time Petri net with priorities in Fig. 10.1. Note: the values of $h(t_{3,2})$ in blue are the values of the clock of the transition $t_{3,2}$ when it is suspended, the values in green correspond to the change of the clock of $t_{3,2}$ when it becomes enabled again from the suspended state, and the values in red illustrate the change of the clock of $t_{3,2}$ while this transition has not been suspended

Obviously, this state means that Task D has been performed for 3 units of time, while Task C is being suspended. Since the resource r_1 becomes idle at this state, Task C may be waked up. As a result, firing $t_{3,1}$ yields the following state

$$(p_6 + p_7 + p_{16}, h(t_{3,2}) = 1 \wedge h(t_{4,2}) = 3)$$

which means that Task C is waked up and its clock starts to work again. Therefore, after scheduling the transition $t_{4,2}$ at this state, the clock value of $t_{3,2}$ becomes $3 = 1 + 2$ because $t_{4,2}$ must be fired after and only after 2 units of time. As a result, the following state is reached:

$$(p_6 + p_9 + p_{10} + p_{16} + r_2, h(t_{3,2}) = 3 \wedge h(t_{6,1}) = 0).$$

Similarly, we can understand other states and their transfers.

Example 10.2 In Fig. 10.2, it seems that $t_{3,1}$ and $t_{5,1}$ are both schedulable at the following state

$$(p_3 + p_4 + p_5 + p_{13} + r_1 + r_2, h(t_{3,1}) = h(t_{4,1}) = h(t_{5,1}) = 0).$$

Actually, the latter can be scheduled but the former cannot, since the latter has a higher priority than the former. It is the reason why there is no $t_{3,1}$-edge outgoing from this state in Fig. 10.2.

Similar to time Petri nets, we can define the *state classes* of a plain time Petri net with priorities as well as its *state class graph*. When computing a *dynamic firing interval* of a schedulable transition, we do not need to take the clock domains of those suspended transitions as constraints, and then use the method introduced in Chap. 8 to do so. Accordingly, when computing a successor of a state class, it is unnecessary to consider the clock domains of those suspended transitions; and after computing a successor, we only need to augment the clock values of the related suspended transitions into the domains.

Figure 10.2 actually shows the state class graph of the plain time Petri net with priorities in Fig. 10.1a. In the next section, we will show that the state class graph of a plain time Petri net with priorities owns some special characteristics.

Question

Try to provide an algorithm of producing the state class graph of a plain time Petri net with priorities.

10.2   Properties of State Class Graphs of Plain Time Petri Nets with Priorities

From Fig. 10.2, we can see that each state class of the state class graph of a plain Petri net is a concrete state, and the dynamic firing interval of a schedulable transition is a fixed rational number but not an interval. Especially, for any two transitions outgoing from the same state class, their dynamic firing intervals at this state class are the same. In what follows, we illustrate these characteristics formally.

Let $R(N^p, s_0)$ and $R(N^p, \pi_0)$ represent the sets of all reachable states and all state classes of a plain time Petri net with priorities (N^p, M_0), respectively.

Theorem 10.1   *For each state class $\pi \in R(N^p, \pi_0)$, we have $\pi \in R(N^p, s_0)$.*

Proof   First, the initial state class is the initial state. Let t be a schedulable transition at the initial state class. Due to

$$\underline{I(t)} = \overline{I(t)},$$

we have that the dynamic firing interval of t at this state class is exactly

$$\underline{I(t)} = \theta_t = \overline{I(t)}.$$

Therefore, after we compute the succeeding state class of this state class according to θ_t, the clock domain of each transition in this successor is still a fixed rational number (but not an interval), i.e., this new state class contains only one state. Similarly, any succeeding state class of this new state class still contains only one state, and so on.   □

Corollary 10.1   *For any state class $\pi \in R(N^p, \pi_0)$ and any transition t schedulable at π, we have $\theta_t \in \mathbb{Q}$.*

Theorem 10.2   *For any state class $\pi \in R(N^p, \pi_0)$ and any two different transitions t_1 and t_2 schedulable at π, we have $\theta_{t_1} = \theta_{t_2}$.*

Proof   Since π contains only one state, we know the clock values of t_1 and t_2 are both a fixed rational number but not an interval. Therefore, we denote them as $h(t_1)$ and $h(t_2)$, and they satisfy

$$\underline{I(t_1)} = h(t_1) + \theta_{t_1} = \overline{I(t_1)} \text{ and } \underline{I(t_2)} = h(t_2) + \theta_{t_2} = \overline{I(t_2)}.$$

We suppose that $\theta_{t_1} \neq \theta_{t_2}$, e.g. $\theta_{t_1} < \theta_{t_2}$. Then, we know that after θ_{t_2} units of time goes from this state, t_2 must be fired; but at this time, the value of the clock of t_1 is $h(t_1) + \theta_{t_2}$ greater than its latest firing time $\overline{I(t_1)}$, which is not permitted. Therefore, we have $\theta_{t_1} = \theta_{t_2}$.   □

Based on the above conclusion as well as the related conclusions (Theorem 8.3 and Corollary 8.1) of time-soundness in Chap. 8, we know that the behaviour of any plain time Petri net with priorities is deterministic.

Corollary 10.2 *A plain time Petri net with priorities is time-sound.*

For convenience, we denote the *state class graph* of a plain time Petri net with priorities as $(R(N^p, \pi_0), E, L)$ where

1. $R(N^p, \pi_0)$ is the set of all reachable state classes,
2. $E \subseteq R(N^p, \pi_0) \times R(N^p, \pi_0)$ is the set of directed edges, and
3. $\forall \pi \in R(N^p, \pi_0): L(\pi) = \theta_t$ if t is schedulable at π, or $L(\pi) = 0$ if no transition is schedulable at π.

In the above definition, the label value of a state class stands for the dynamic firing interval (that is actually a rational number) of all schedulable transitions at this state class.

10.3 Verifying TCTL Based on State Class Graphs of Plain Time Petri Nets with Priorities

Given a TCTL formula describing a property of a plain time Petri net with priorities, we can utilise the model checking algorithms of TCTL in Chap. 9 to verify it. But we should be careful for the clock values of those suspended transitions. When a transition is suspended at a state class, its clock value should not be taken into account. On the other hand, because of the characteristics of the state class graph of a plain time Petri net with priorities, it becomes easy to compute the guided clock domains and guided dynamic firing intervals. We do not introduce them here, and they are left to readers.

Sometimes, we may also use the backward algorithms similar to those in Chap. 5 to compute the satisfiable state set of a TCTL formula due to those characteristics introduced in the above section. What we additionally need to do is to accumulate the time value of an edge (i.e., the dynamic firing interval of a state class) and to decide if a sum has been in the time interval of a formula. However, this method does not work for some plain time Petri nets with priorities; later, we will show such a case that the following algorithms are unsuitable for.

Algorithm 10.1 describes the framework of recursively computing the satisfiable state set of a TCTL formula, that is similar to Algorithm 9.1. Next, we illustrate the computations of an $\mathbf{EU_I}$ formula and an $\mathbf{AU_I}$ formula, respectively, that are involved in Lines 10 and 12 in Algorithm 10.1.

Algorithm 10.1: SAT-PT(ϕ): Computing the Satisfiable State Set of a TCTL Formula ϕ Based on a Plain Time Petri Net with Priorities.

Input: A state class graph $(R(N^p, M_0), E, L)$ and a TCTL formula ϕ;
Output: The satisfiable state set **SAT-PT(ϕ)**;

1 **if** $\phi = $ **true then**
2 | **return** $R(N^p, \pi_0)$;

3 **if** $\phi = p$ **then**
4 | **return** $\{\pi = (M, C) \in R(N^p, \pi_0) \mid M(p) \geq 1\}$;

5 **if** $\phi = \neg\phi_1$ **then**
6 | **return** $R(N^p, \pi_0) \setminus$ **SAT-PT(ϕ_1)**;

7 **if** $\phi = \phi_1 \wedge \phi_2$ **then**
8 | **return SAT-PT(ϕ_1) $\cap$ SAT-PT(ϕ_2)**;

9 **if** $\phi = \mathbf{E}(\phi_1 \mathbf{U_I} \phi_2)$ **then**
10 | **return SAT-PT$_{\mathbf{EU}}(\phi_1, \phi_2, \mathbf{I})$**;

11 **if** $\phi = \mathbf{A}(\phi_1 \mathbf{U_I} \phi_2)$ **then**
12 | **return SAT-PT$_{\mathbf{AU}}(\phi_1, \phi_2, \mathbf{I})$**;

10.3.1 Computing Satisfiable State Set for EU_I Based on Backward Exploration

To compute the satisfiable state set of a formula $\mathbf{E}(\phi_1 \mathbf{U_I} \phi_2)$, we construct an ordered pair $(\pi, time)$ recording the time from the node π (satisfying ϕ_1) to a target node (satisfying ϕ_2). In consequence, a node possibly corresponds to multiple such ordered pairs in the computing procedure since there are possibly multiple such paths starting at this node.

We compute the satisfiable state set for each element in Y. Starting at such a node, we travel backward its predecessors as well as its ancestors that are all in X. If the time length is not beyond the upper bound of $\mathbf{I}$, then this traversal will go on. Certainly, if the time sum accumulated at some node is in $\mathbf{I}$, then this node is satisfiable. Obviously, Algorithm 10.2 is suitable for the case of $\bar{\mathbf{I}} \neq \infty$. The case of $\bar{\mathbf{I}} = \infty$ is left to readers.

Example 10.3 We consider the following formula in Fig. 10.2:

$$\mathbf{E}\Big((p_4 \wedge p_5)\mathbf{U}_{[0..1]}\big((p_5 \wedge r_1) \vee (p_4 \wedge r_2)\big)\Big).$$

Since the state class

$$\pi_1 = (p_3 + p_4 + p_{15} + r_2, h(t_{4,1}) = h(t'_{5,2}) = 0 \wedge h(t_{3,2}) = 1)$$

satisfies $(p_5 \wedge r_1) \vee (p_4 \wedge r_2)$ and $0 \in [0..1]$, it is satisfiable. Its predecessor

$$\pi_2 = (p_4 + p_5 + p_6 + p_{13} + r_2, h(t_{4,1}) = h(t'_{5,1}) = 0 \wedge h(t_{3,2}) = 1)$$

Algorithm 10.2: SAT-PT$_{EU}(\phi_1, \phi_2, \mathbf{I})$: Computing the Satisfiable State Set of $\mathbf{E}(\phi_1 \ \mathbf{U_I} \ \phi_2)$ Based on a Plain Time Petri Net with Priorities.

Input: A state class graph $(R(N^P, M_0), E, L)$, two formulae ϕ_1 and ϕ_2, and an interval $\mathbf{I}$;
Output: The satisfiable set **SAT-PT$_{EU}(\phi_1, \phi_2, \mathbf{I})$**;

1   $X \leftarrow$ **SAT-PT**(ϕ_1);
2   $Y \leftarrow$ **SAT-PT**(ϕ_2);
3   $Z \leftarrow \emptyset$;
4   **for** *each* $\pi \in Y$ **do**
5      $X_1 \leftarrow \{(\pi, 0)\}$;
6   **while** $X_1 \neq \emptyset$ **do**
7   Take an element out of X_1, which is denoted as $(\pi_1, time)$;
8   **if** $time \leq \overline{\mathbf{I}}$ **then**
9   **if** $time \geq \underline{\mathbf{I}}$ **then**
10              $Z \leftarrow Z \cup \{\pi_1\}$;
11            $X_1 \leftarrow X_1 \cup \{(\pi_2, time + L(\pi_2)) \mid \pi_2 \in X, (\pi_2, \pi_1) \in E\}$;

12   **return** Z;

---

satisfies $p_4 \wedge p_5$ and $L(\pi_2) + 0 = 0 + 0 = 0 \in [0..1]$, π_2 is also satisfiable. The predecessor of π_2 is

$$\pi_3 = (p_4 + p_5 + p_6 + p_{12}, h(t_{2,2}) = 4 \wedge h(t_{3,2}) = 0).$$

This state class also satisfies $p_4 \wedge p_5$ and $L(\pi_3) + 0 = 1 + 0 = 1 \in [0..1]$, it also satisfiable. Continually, we can find the next predecessor

$$\pi_4 = (p_3 + p_4 + p_5 + p_{12} + r_1, h(t_{2,2}) = 4 \wedge h(t_{3,1}) = 0)$$

is also satisfiable. In addition to the above satisfiable path, we also find the following two satisfiable ones:

$$(p_3 + p_4 + p_5 + p_{12} + r_1, h(t_{2,2}) = 5 \wedge h(t_{3,1}) = 0)$$

$$\downarrow (0, t_{2,2})$$

$$(p_3 + p_4 + p_5 + p_{13} + r_1 + r_2, h(t_{3,1}) = h(t_{4,1}) = h(t_{5,1}) = 0)$$

$$\downarrow (0, t_{4,1})$$

$$(p_3 + p_5 + p_7 + p_{13} + r_1, h(t_{3,1}) = h(t_{4,2}) = h(t_{5,1}) = 0)$$

and

$$(p_3 + p_4 + p_5 + p_{12} + r_1, h(t_{2,2}) = 5 \wedge h(t_{3,1}) = 0)$$

$$\downarrow (0, t_{2,2})$$

$$(p_3 + p_4 + p_5 + p_{13} + r_1 + r_2, h(t_{3,1}) = h(t_{4,1}) = h(t_{5,1}) = 0)$$

$$\downarrow (0, t_{5,1})$$

$$(p_3 + p_4 + p_{14} + r_2, h(t_{4,1}) = h(t_{5,2}) = 0).$$

For the first path of this example, if we change the time interval of the formula into [0..0.8], only π_1 and π_2 are found in this path. Actually, some states in π_3 still satisfy the modified formula, e.g. the following state subclass

$$(p_4 + p_5 + p_6 + p_{12}, 4.2 \le h(t_{2,2}) \le 5 \land 0.2 \le h(t_{3,2}) \le 1 \land h(t_{2,2}) - h(t_{3,2}) = 4).$$

The reason lies in Line 8 of Algorithm 10.2, which handles the upper-bound problem too simply to compute the above state subclass. When the time length of an existing path is exactly equal to the upper bound of $\mathbf{I}$ and accumulating the dynamic firing interval of the current explored state class results in crossing the upper bound, it has no problem to abandoning this explored state class. However, if the time length of an existing path has not reached the upper bound of $\mathbf{I}$ but it will cross this upper bound after accumulating the dynamic firing interval of the current explored state class, abandoning this explored state class can results in a problem as shown in the above case. Hence, there is a limitation to the application of this algorithm.

10.3.2   Computing Satisfiable State Set for AU$_I$ Based on Backward Exploration

On the basis of the satisfiable set of $\mathbf{E}(\phi_1 \, \mathbf{U_I} \, \phi_2)$, we easily compute the satisfiable state set of $\mathbf{A}(\phi_1 \, \mathbf{U_I} \, \phi_2)$. If some element in the satisfiable set of $\mathbf{E}(\phi_1 \, \mathbf{U_I} \, \phi_2)$ satisfies ϕ_1 and has a successor not belonging to this satisfiable state set, then this element and all its ancestors do not satisfy $\mathbf{A}(\phi_1 \, \mathbf{U_I} \, \phi_2)$. Therefore, we only need to remove such elements out of $\mathbf{E}(\phi_1 \, \mathbf{U_I} \, \phi_2)$, and the remainders form the satisfiable set of $\mathbf{A}(\phi_1 \, \mathbf{U_I} \, \phi_2)$. It is worthy noting another case that some element in the satisfiable set of $\mathbf{E}(\phi_1 \, \mathbf{U_I} \, \phi_2)$ has a successor not belonging to this satisfiable set but satisfies ϕ_2, this means that this element is the last node of satisfiable paths; hence, such an element should not be removed. Algorithm 10.3 describes this procedure.

Example 10.4 We consider the following formula in Fig. 10.2:

$$\mathbf{A}\Big((p_4 \land p_5)\mathbf{U}_{[0..1]}\big((p_5 \land r_1) \lor (p_4 \land r_2)\big)\Big).$$

Algorithm 10.3: SAT-PT$_{AU}$(ϕ_1, ϕ_2, I): Computing the Satisfiable State Set of A(ϕ_1 U$_I$ ϕ_2) Based on a Plain Time Petri Net with Priorities.

Input: A state class graph ($R(N^t, \pi_0)$, E, L), two formulae ϕ_1 and ϕ_2, and an interval **I**;
Output: The satisfiable set **SAT-T$_{AU}$**(ϕ_1, ϕ_2, **I**);
1 $X \leftarrow$ **SAT-PT$_{EU}$**(ϕ_1, ϕ_2, **I**);
2 $Y \leftarrow$ **SAT-PT**(ϕ_1);
3 **for** *each* $\pi \in X$ **do**
4   **if** $\pi \in Y$ *and there is* $\pi' \notin X$ *such that* $(\pi, \pi') \in E$ **then**
5   Remove π and all its ancestors out of X;

6 **return** X;

In Example 10.3, we have obtained three satisfiable paths for the formula

$$\mathbf{E}\Big((p_4 \wedge p_5)\mathbf{U}_{[0..1]}\big((p_5 \wedge r_1) \vee (p_4 \wedge r_2)\big)\Big).$$

But we find that for the following state class of the first path:

$$\pi_2 = (p_4 + p_5 + p_6 + p_{13} + r_2, h(t_{4,1}) = h(t'_{5,1}) = 0 \wedge h(t_{3,2}) = 1),$$

its successor

$$(p_5 + p_6 + p_7 + p_{13}, h(t_{4,2}) = h(t'_{5,1}) = 0 \wedge h(t_{3,2}) = 1)$$

unsatisfies

$$\mathbf{E}\Big((p_4 \wedge p_5)\mathbf{U}_{[0..1]}\big((p_5 \wedge r_1) \vee (p_4 \wedge r_2)\big)\Big).$$

Therefore, this path does not satisfy

$$\mathbf{A}\Big((p_4 \wedge p_5)\mathbf{U}_{[0..1]}\big((p_5 \wedge r_1) \vee (p_4 \wedge r_2)\big)\Big).$$

Other two paths in Example 10.3 satisfy this formula.

10.4 Application

This application example comes from a communication company in China. According to the security requirement of this company, we do not describe the concrete detail of every task here.

System Description and Model

There are six tasks in this system and their dependent relation, performing times, and priorities are shown in Fig. 10.3a. To meet the real-time demand, every task is tied to a fixed CPU, e.g. Tasks A, B, and F use CPU r_1 and Tasks C, D, and E use CPU r_2.

Tasks A and D are driven periodically, and the periods are 100 and 50 units of time, respectively. In the Petri net model in Fig. 10.3b, we use transitions $t_{0,1}$ and $t_{0,2}$ to model the two periodically-driven events. After Tasks A and D are finished, Tasks B, E, and F are driven. Task C is driven only after Tasks B and E are both finished.

Although Task B has a higher priority than Task A, it cannot happen that Task B preempts r_1 from Task A, since Task B can immediately be scheduled to perform once Task A releases r_1. Similarly, Task E does not preempt r_2 from Task D.

It looks possible that Task A or B preempts r_1 from Task F. We may add the simulation of the two preempting cases into our model, but they do not happen actually since the related transitions are disabled forever. Therefore, we do not need to consider them in our model.

Task C probably preempts r_2 from Task D or E. It can be checked that only the case of Task C preempting r_2 from Task E can take place. Hence, we use the transition $t_{3,3}$ to model this preemption.

This system takes an *overwrite* policy. After a task (e.g. Task E) is finished and drives another task (e.g. Task C), the latter has not been performed, but the former is performed again and drives the latter again. At this time, the previous message of

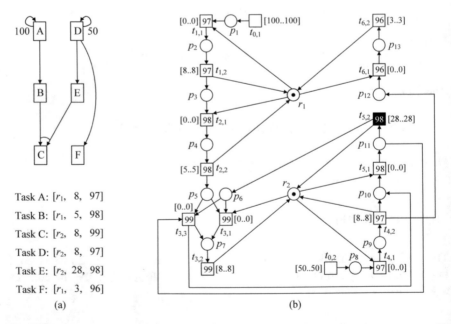

Task A: $[r_1,\ 8,\ 97]$

Task B: $[r_1,\ 5,\ 98]$

Task C: $[r_2,\ 8,\ 99]$

Task D: $[r_2,\ 8,\ 97]$

Task E: $[r_2,\ 28,\ 98]$

Task F: $[r_1,\ 3,\ 96]$

(a)

(b)

Fig. 10.3 a The task diagram in a communication system, and **b** the plain time Petri net with priorities modelling this system

driving this task is covered by the current one. It is easily simulated in a Petri net model through a simple modification of the firing rule: when a place has a token and another token is produced into this place, then the number of tokens in this place maintains 1 (i.e., $1 + 1 = 1$). In [1], we use such a firing rule.

Figure 10.4 illustrates the state class graph of this net system. From it we can see that the class state

$$(2p_6 + r_1 + r_2, h(t_{0,1}) = 86 \land h(t_{0,2}) = 36)$$

covers the previous one

$$(p_6 + r_1 + r_2, h(t_{0,1}) = 86 \land h(t_{0,2}) = 36).$$

If we use the overwrite policy, the former becomes the latter and thus this state class graph becomes finite.

Design Requirements and Analysis

The following design requirements are concerned:

1. Is there a task never scheduled or never finished?
2. Can a task always be finished in a deadline once it is driven?
3. How long is the maximum idle time of a CPU?

The first one is to check if there is a starvation (unfairness). We can use a CTL formula or a TCTL one to represent if there is a computation path in which some task has never been scheduled in future. For example of Task F, we may construct the following formulae:

$$\mathbf{EF}\left(\mathbf{EG}\neg p_{13}\right) \text{ or } \mathbf{EF}_{[0..\infty)}\left(\mathbf{EG}_{[0..\infty)}\neg p_{13}\right).$$

The two formulae do not hold at the initial state, which means that Task F can always be scheduled.

For verifying the second requirement w.r.t. some task, we may augment an additional place and an additional transition into the model. For example, when we plan to check if Task E can be finished in its deadline (e.g. 36 units of time), we add a place p_0 and a transition t_0 into the model in Fig. 10.3b such that:

$$^\bullet p_0 = \{t_{5,2}\} \land p_0^\bullet = \{t_0\} \land {}^\bullet t_0 = \{p_0\} \land t_0^\bullet = \emptyset \land I(t_0) = [0..0].$$

Then, we may construct the following formula

$$\mathbf{AG}_{[0..\infty)}\left(p_{10} \Rightarrow \mathbf{AF}_{[0..36]} p_0\right),$$

where p_{10} means Task E is driven and p_0 means it is finished. This formula holds.

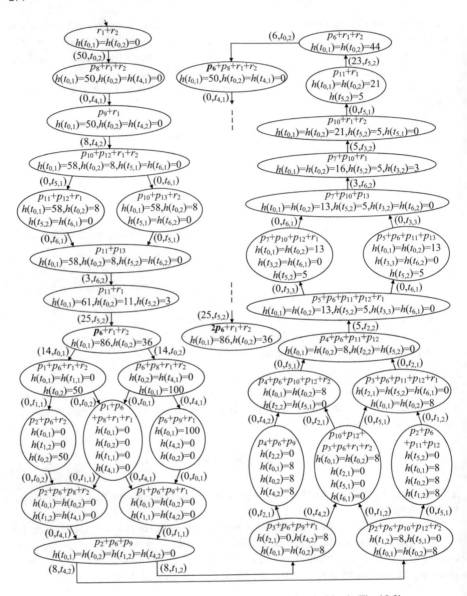

Fig. 10.4 The state class graph of the plain time Petri net with priorities in Fig. 10.3b

The last requirement is about the performance of this system. We may utilise a TCTL formula to probe such a maximum value through setting up different time intervals of the formula. For example of r_1, we may construct the following TCTL formula

$$\mathbf{EF}_{[200..\infty)}\Big(\mathbf{E}(r_1\mathbf{U}_{[x..x+1]}\neg r_1)\Big).$$

When we assign x a value and the related formula holds, we can continue to enlarge the value of x and probe it again. It is possible that this kind of method cannot compute an accurate value, but it can provide a very approximate one. Here, the interval $[200..\infty)$ of $\mathbf{EF}$ means we consider the behaviours of this system after it runs 200 units of time. When we probe the case of $[42..43]$, this formula still holds, but it does not hold for $[43..44]$. Hence, the maximum idle time of r_1 is about 43 units of time, while it is actually 42.

10.5 Summery and Further Reading

This chapter introduces plain time Petri nets with priorities, modelling some real-time resource allocation system in which resources such as CPU may be preempted. Their state class graphs are of deterministic characteristics, bringing convenience for the analysis to them. There are some work considering priorities in time Petri nets [2–7], but it is not convenient for them to model some complex cases, e.g. a task can be performed on multiple different resources. There are lots of work [8–19] on formally modelling and analysing real-time systems under considering scheduling policies with or without priorities. It is a challenge how to model the real-time system in which the priority of a task may be changed in the running, and it is an interesting topic how to use logical formulae to compute a system's performance.

References

1. He, L.F., Liu, G.J.: Time-point-interval prioritized time Petri nets modelling real-time systems and TCTL checking. *Journal of Software* 33: 2947–2963 (2022) (in Chinese, https://doi.org/10.13328/j.cnki.jos.006607)
2. Roux, O.H., Déplanche, A.M.: A t-time Petri net extension for real time-task scheduling modeling. *European Journal of Automation* 36: 973–987 (2002)
3. Bucci, G., Fedeli, A., Sassoli, L., Vicario, E.: Timed state space analysis of real-time preemptive systems. *IEEE Transactions on Software Engineering* 30: 97–111 (2004)
4. Roux, O.H., Lime, D.: Time Petri nets with inhibitor hyperarcs: Formal semantics and state space computation. In: *the International Conference on Application and Theory of Petri Nets*, pp. 371–390 (2004)
5. Berthomieu, B., Lime, D., Roux, O.H., Vernadat, F.: Reachability problems and abstract state spaces for time Petri nets with stopwatches. *Discrete Event Dynamic Systems* 17: 133–158 (2007)

6. Battyányi, P., Vaszil, G.: Description of membrane systems with time Petri nets: promoters/inhibitors, membrane dissolution, and priorities. *Journal of Membrane Computing* 2: 341–354 (2020)
7. Matsuoka, R., Nakata, A.: Deadline assignment optimization method using extended time Petri nets for real-time multitask distributed systems sharing processors with EDF scheduling. *IEICE Technical Report* 117: 53–58 (2018)
8. Bolton, M.L., Edworthy, J., Boyd, A.D.: A formal analysis of masking between reserved alarm sounds of the IEC 60601-1-8 international medical alarm standard. In: *Proceedings of the Human Factors and Ergonomics Society Annual Meeting*, pp. 523–527 (2018)
9. Soumia, E., Ismail, A., Mohammed, S.: Formal method for the rapid verification of SystemC embedded components. In: *the 5th International Conference on Optimization and Applications*, pp. 1–5 (2019)
10. Yuan, Z., Chen, X., Liu, J., Yu, Y., Sun, H., Zhou, T., Jin, Z.: Simplifying the formal verification of safety requirements in zone controllers through problem frames and constraint-based projection. *IEEE Transactions on Intelligent Transportation Systems* 19: 3517–3528 (2018)
11. Pimkote, A., Vatanawood, W.: Simulation of preemptive scheduling of the independent tasks using timed automata. In: *the 10th International Conference on Software and Computer Applications*, pp. 7–13 (2021)
12. Hu, Y., Wang, J.C., Liu, G.J.: Resource-oriented timed workflow nets and simulation tool design. In: *2021 International Conference on Cyber-Physical Social Intelligence*, pp. 1–6 (2021)
13. Qi, L., Luan, W., Liu, G.J., Lu, X.S., Guo, X.W.: A Petri net based traffic rerouting system by adopting traffic lights and dynamic message signs. In: *2020 IEEE International Conference on Networking, Sensing and Control*, pp. 1–6 (2020)
14. Ding, Z., Zhou, Y., Zhou, M.C.: A polynomial algorithm to performance analysis of concurrent systems via Petri nets and ordinary differential equations. *IEEE Transactions on Automation Science and Engineering* 12: 295–308 (2013)
15. Pan, L., Yang, B., Jiang, J., Zhou, M.C.: A time Petri net with relaxed mixed semantics for schedulability analysis of flexible manufacturing systems. *IEEE Access* 8: 46480–46492 (2020)
16. Zhou, J.N., Wang, J.C., Wang, J.: A simulation engine for stochastic timed Petri nets and application to emergency healthcare systems. *IEEE/CAA Journal of Automatica Sinica* 6: 969–980 (2019)
17. Yang, F., Wu, N.Q., Qiao, Y., Su, R.: Polynomial approach to optimal one-wafer cyclic scheduling of treelike hybrid multi-cluster tools via Petri nets. *IEEE/CAA Journal of Automatica Sinica* 5: 270–280 (2017)
18. Lefebvre, D., Basile, F.: An approach based on timed Petri nets and tree encoding to implement search algorithms for a class of scheduling problems. *Information Sciences* 559: 314–335 (2021)
19. Lee, J.H., Kim, H.J.: Reinforcement learning for robotic flow shop scheduling with processing time variations. *International Journal of Production Research* 60: 2346–2368 (2022)

Index

Printed in the United States
by Baker & Taylor Publisher Services